Phonics and Spelling

Ages 10–11

Julie Crimmins-Crocker

Published by Collins
An imprint of HarperCollins*Publishers*
77–85 Fulham Palace Road
Hammersmith
London
W6 8JB

Browse the complete Collins catalogue at
www.collinseducation.com

First published in 2007 by Folens Limited.

ISBN-13: 978-0-00-745240-8

British Library Cataloguing in Publication Data
A catalogue record for this publication is available from the British Library.

Managing editor: Joanne Mitchell
Layout artist: Neil Hawkins, ndesign
Illustrations: JB Illustrations, Helen Jackson and Nicola Pearce of SGA and Leonie Shearing c/o Lucas Alexander Whitley.
Cover design for this edition: Julie Martin
Design and layout for this edition: Linda Miles, Lodestone Publishing
Printed and bound in China.

Contents

This contents list provides an overview of the learning objectives of each puzzle page.

How to complete the puzzles 3
What's next? 5
Words with **ACE** and **ASE** 6
Words with **AGE** 7
Words with **AID** and **ADE** 8
Words with **AIL** and **ALE** 10
Words with **AKE** 12
Words with **AME** and **AIM** 13
Words with **AIN** and **ANE** 14
Words with **AY** 15
Words with **ATE** 16
Words with **EED**, **EAD** and **EDE** 17
Words with **ANGE** 18
Words with **AIR** and **ARE** 20
Words with **EAK** and **EEK** 22
Words with **EAL** and **EEL** 23
Words with **EAP** and **EEP** 24
Words with **EER** and **EAR** 25
Words with **EAT** and **EET** 26
Words with **ICE** 28
Words with **IDE** and **IED** 29
Words with **IGHT** and **ITE** 30
Words with **ILE** and **IAL** 32
Words with **IRE** 33
Words with **INE** 34
Words with **OKE** 36
Words with **OPE** 38
Words with **OTE** and **OAT** 39
Words with **OOD** and **UDE** 40
Words with **OON** 42
Words with **OOK** 44
Words with **OW** 45
Words with **OR** and **ORE** 46
Words that end in **ER** 48
Words that end in **AR** or **OR** 49
Silent **E** 50
Long **A** phoneme 52
Long **E** phoneme 54
Long **I** phoneme 56
Long **O** phoneme 58
Long **U** phoneme 60
Plural nouns 62
Verbs with **ED** and **ING** 64
Hard and soft **C** and **G** 65
Words with **GH** and **GHT** 66
I before **E** except after **C** 67
Suffixes – changing nouns and adjectives into verbs 68
Changing verbs into nouns 69
Adjectives and adverbs 70
Suffixes 71
Suffixes revision 72
Prefixes 74
Prefixes revision 76
Misspelt words with unstressed vowels 78
Commonly misspelt words or easily confused spellings 81
Answers 84

How to complete the puzzles

1. Read the title and the instructions for each activity very carefully.
2. For each activity, start with the simplest clues first.
3. Crossword and wordsearch clues have numbers at the end of each clue to tell you how many letters there are in the word.
4. If there is a word bank for you to refer to, check that your answers are in the list. Cross out the words in the word bank as you complete the clue.
5. Use a sharp pencil at first to write down all your answers (just in case you need to change them).
6. When you are sure your answers are correct, write them in pen or use a highlighter pen for the wordsearches.
7. In the crosswords, write in CAPITAL LETTERS. This will make your answers easier to read.
8. Use a dictionary and thesaurus to help you spell and find your answers.
9. After each puzzle go to 'What's next' (page 5) and cross off the completed activity.

Remember what these useful words mean:

Synonym: the same or similar meaning, for example, ***big* – *large***.

Antonym: the opposite meaning, for example, ***big* – *small***.

Anagram: the word is muddled up, for example, ***GREAL* – *LARGE***.

Informal: the word is simple or slang, for example, ***rabbit* – *bunny***.

Verbs are action and 'doing' words, for example, ***run***, ***talk*** and ***think***.

Nouns are naming words, for example, ***pen***, ***hat***, ***apple*** and ***school***.

Adjectives are describing words, for example, ***small*** *bird*.

Adverbs add more information to verbs, for example, *He ran* ***quickly***.

Phoneme: a letter or letters that create a single sound when said aloud, for example, **TH** and **OO**.

Letter string: a collection of phonemes, for example, **ELL**.

Vowels are the letters **a**, **e**, **i**, **o** and **u**.

Consonants are the letters of the alphabet which are not vowels.

What's next?

Use the answers to any of the puzzles to complete the following activities. Write down which puzzle you have completed and the date you did it.

	Activity	Puzzle title	Date
★ 1	Sort the answers into **alphabetical order**. Put them in a list.		
★ 2	Put the answers into **sentences** (10 sentences minimum). You can use one word per sentence or include as many words as you like in each sentence.		
★ 3	Put the answers into sentences that are questions. For example, *Where did my cat go?*		
★ 4	Put the words into sentences that are instructions. For example, *Look after my cat when I am away.*		
★ 5	Put the words in the word bank into a story or piece of writing.		
★ 6	Write at least ten more of the same type of words.		
★ 7	Find **synonyms** for the words and write them down in pairs or groups. For example, *big – large, massive.*		
★ 8	Find **antonyms** for the words and write them down in pairs or groups. For example, *big – small, tiny, minute.*		
★ 9	Find **rhymes** for the words and list them. For example, *ink – sink* and *bellow – yellow, mellow* and *fellow.*		
★ 10	Sort the answers into groups. For example, verbs, nouns, adjectives, adverbs, number of syllables, rhyming words and so on.		
★ 11	Write new clues for the answers to a crossword, wordsearch or puzzle or create a totally new puzzle.		

Words with ACE and ASE

The long **A** phoneme is heard in words containing the letter strings **ACE** or **ASE**, for example, *place, space* and *case*. When **ED** is added to verbs ending in **ACE** and **ASE**, the letter strings remain the same, for example, *race – raced*. When **ING** and **Y** are added, the **E** is removed, for example, *race – racing*.

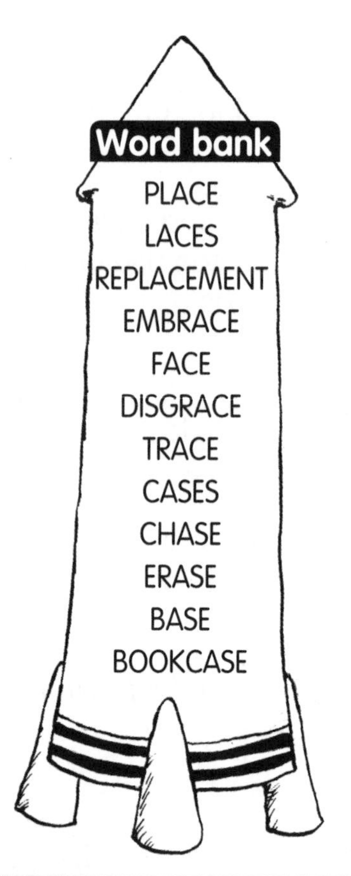

The answers in this crossword have the letter strings **ACE** or **ASE** and a long **A** phoneme.

ACROSS	**DOWN**
4. A location or spot. (5)	**1.** Tie these to keep your shoes on (5)
6. To rub out or remove (5)	**2.** A substitute or understudy (11)
9. Pack these when you go on holiday (5)	**3.** Shame or dishonour (8)
10. The front of your head with your features (4)	**5.** To hunt or pursue someone or something (5)
11. To hug or clasp into your arms (7)	**7.** A place where reading material is stored on shelves (8)
12. To copy exactly through thin paper (5)	**8.** The bottom or foundation (4)

Sort these words into two groups in the table below.

deface necklace erase solace furnace staircase
disgrace terrace misplace replace purchase palace

ASE and ACE – long A phoneme		**ASE and ACE – short A phoneme**	

Words with AGE

The letter string **AGE** is usually pronounced with a long **A** phoneme, for example, *rage* and *page*. However, when **AGE** is added as a **suffix** it is often pronounced more like **IDGE**, for example, *damage, shrinkage* and *anchorage*. When **ED** is added to verbs ending in **AGE**, the letter string remains the same, for example, *cage – caged*. When **ING** is added, the **E** is removed, for example, *damage – damaging.*

These words and definitions have got muddled up. Write the correct word next to each definition. The first one has been done for you.

Word	Definition	Correct word
ANCHORAGE	A very wise man and an aromatic herb.	SAGE
BARRAGE	The mechanical power of a lever or influence or advantage used for a purpose.	
STEERAGE	Bravery.	
LEVERAGE	A mooring place for boats.	
HOSTAGE	To make angry.	
COURAGE	Part of a passenger ship with the lowest fares.	
WAGER	Heavy artillery fire.	
~~SAGE~~	A person held prisoner for ransom, or as security.	
WAGES	A bet.	
ENRAGE	Payment for work done.	

Unjumble the anagrams in brackets and write the missing **AGE** words in the sentences.

1. The severe storm caused a lot of __________________ to trees and buildings. (GAMEAD)
2. The __________________ lions were roaring and trying to escape. (DACEG)
3. You usually have to pay the __________________ when you buy something by mail order. (GESTOPA)
4. The teacher said, 'This paint is very expensive so I do not want any __________________'. (SAWGATE)
5. When you are in a foreign country you have to get used to their money which will have different notes and __________________. (GOANICE)
6. The actors strutted around on the __________________. (GATES)
7. Because of gradual __________________ the water tank ran dry. (KEAGALE)
8. If you are very tall, you may have an __________________ when you play basketball. (GATEDANAV)

Words with AID and ADE

The long **A** phoneme is heard in words containing **AID** or **ADE**, for example, *afraid* and *trade*. The **AID** letter string often occurs when verbs ending in **AY** are changed into the past tense, for example, *pay – paid*. Some verbs that end in **AY** when changed to the past tense do not lose the **Y**, for example, *play – played*. They have the letter string **AYED**. When **ED** is added to verbs ending in **ADE**, the letter string remains the same, for example, *fade – faded*. When **ING** and **Y** are added, the **E** is removed, for example, *fade – fading* and *shade – shady*.

Solve the clues and then circle or highlight the answers in the wordsearch. All the answers contain the letter strings **AID**, **ADE** or **AYED**.

1. A jam made with oranges and eaten on toast (9)
2. Part of a knife that you cut with (5)
3. Ten years (6)
4. To avoid or elude (5)
5. To become lighter and lose colour (4)
6. To enter by force, like an army might do to another country (6)
7. A blue-green colour and an ornamental semi-precious stone (4)
8. A garden tool used for digging (5)
9. To buy and sell things (5)
10. Scared or frightened (6)
11. An attack or invasion (4)
12. A mythical female sea creature with a human body and a fish's tale (7)
13. The past tense of pay (4)
14. Sedate (5)
15. Remained and didn't leave (6)

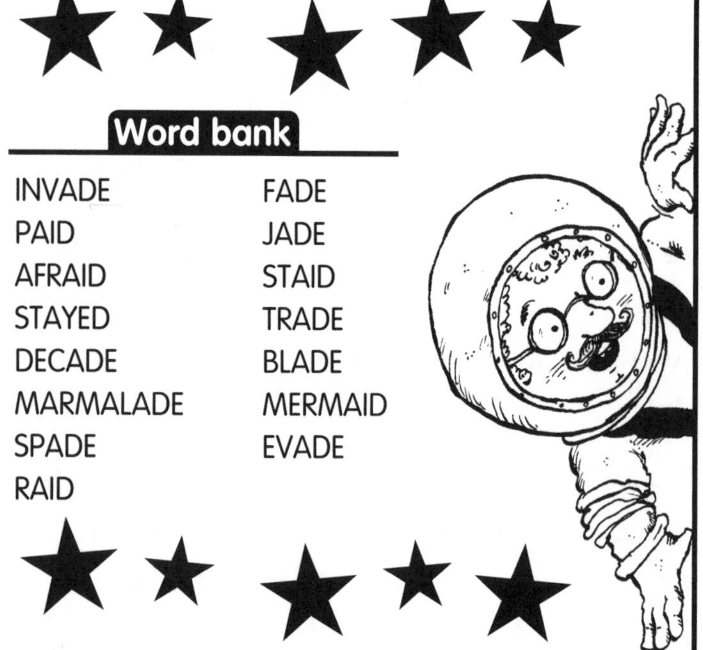

Word bank

INVADE
PAID
AFRAID
STAYED
DECADE
MARMALADE
SPADE
RAID
FADE
JADE
STAID
TRADE
BLADE
MERMAID
EVADE

R	M	E	R	M	A	I	D	M	D	N
E	K	R	S	M	M	I	L	E	Z	E
V	S	A	P	D	A	A	D	B	D	T
A	T	I	A	P	E	A	L	A	H	E
D	A	D	D	M	L	C	F	C	D	P
E	I	H	E	A	L	D	A	A	G	L
J	D	H	M	L	E	Q	R	D	L	O
F	A	R	R	Y	L	T	F	Y	E	F
Z	A	D	A	J	I	N	V	A	D	E
M	L	T	E	N	A	F	R	A	I	D
X	S	T	G	N	R	B	L	A	D	E

Words with AID and ADE

Sort these **AID** and **ADE** words into the table below. If they belong in more then one group, write them in more than one column.

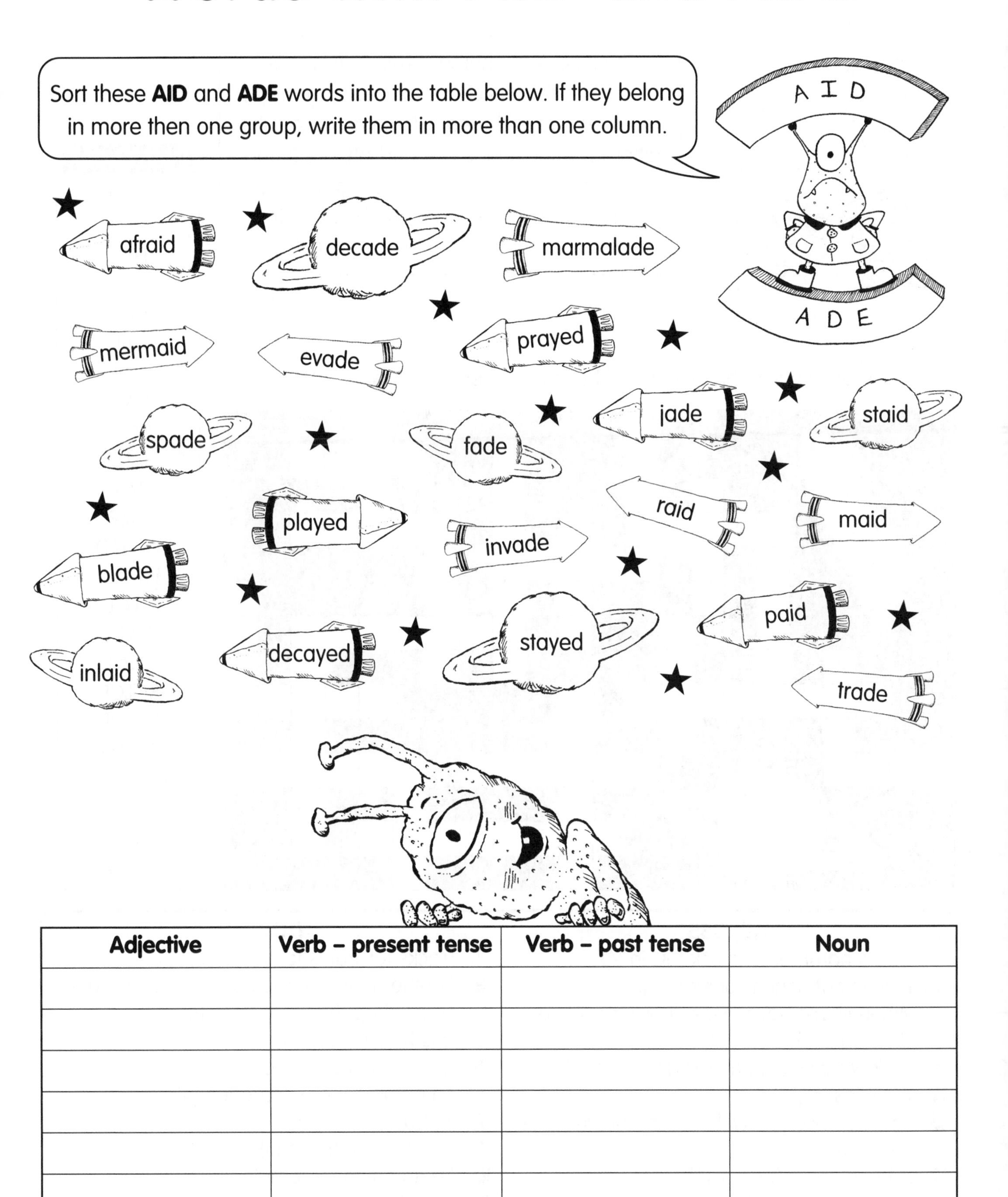

Adjective	Verb – present tense	Verb – past tense	Noun

Words with AIL and ALE

The long **A** phoneme is heard in words containing **AIL** or **ALE**, for example, *ailment, snail, impale* and *whale*. In some cases, the different letter strings **ALE** and **AIL** are found in words which are homophones and the spelling helps us to work out the meaning of the word, for example, *tail – tale*.

Word bank

FAILED
GALE
HAIL
MAIL
MALE
NAIL
PAIL
PALE
SAILS
SALES
TAIL
TALE
WAILED
WHALE
FRAIL
RAILS
SNAIL
SAILOR
TAILOR
AILMENT
RAILWAY
TRAILER
HANDRAIL
PONYTAIL

The answers in this crossword have the letter strings **AIL** or **ALE**.

ACROSS

4. Use a hammer to knock this in (4)
6. Fragile not healthy or strong (5)
8. When shops have these, there are lots of discounts (5)
9. A slow moving creature with a shell (5)
12. An illness (7)
15. This will help you walk up and down stairs safely (8)
17. A large sea mammal (6)
18. Not dark, washed out (4)
19. Cried or moaned in sorrow or pain (6)
20. A person who makes men's clothes (6)
21. Long hair tied at the back of the head (8)

DOWN

1. Tracks for trains to run on (5)
2. Another word for the bucket that Jack and Jill went up the hill to fetch (4)
3. This type of station has trains in it (7)
5. Another word for the post (4)
6. An antonym for succeeded (4)
7. A strong wind (4)
8. Yachts have these (5)
10. An antonym for female (4)
11. Hard frozen raindrops (4)
13. You can tow this behind a car (7)
14. A story (4)
16. A seaman or someone in the navy (6)
20. A dog wags this when it is happy (4)

Words with AIL and ALE

Sort these words into groups in the table below. If a word belongs in more than one group, write it in both.

Noun – singular		Noun – plural	Verb – present tense
		Adjective	**Verb – past tense**

Words with AKE

The long **A** phoneme is heard in words with the letter string **AKE** (with a silent **E**), for example, *sake*. When **ED** is added to verbs ending in **AKE**, the letter string can still be seen, for example, *baked*. When **ING** and **Y** are added, the **E** is removed, for example, *bake – baking* and *shake – shaky*.

Write the missing **AKE** words in these sentences.

bake cake fake lake mistakes awake
drake brakes earthquake undertaker snowflakes

1. I am going to ______________ a chocolate ______________ for tea. (4)
2. The antique was not real; it was a ______________. (4)
3. We went rowing on the ______________. (4)
4. I made lots of ______________ in my spelling test and got lots of words wrong. (8)
5. During the thunderstorm I couldn't get to sleep and I lay ______________ in bed all night. (5)
6. A male duck is called a ______________. (5)
7. I heard the squeal of the car's ______________ before the accident. (6)
8. The tremors of the ______________ measured 7.5 on the Richter scale. (10)
9. A person who arranges funerals is a called an ______________. (10)
10. The pretty white ______________ covered the icy ground like a lacy blanket. (10)

Sort these words into groups: 'Nouns', 'Verbs', 'Adjectives' and 'Adverbs'. Some of them may belong in more than one group, for example, *I am going to **fake** an illness to get off school.* (verb) and *The picture was a **fake**.* (noun).

Jake bake cake fake lake awake drake milkshake
shaky take earthquake undertaker snowflakes stake undertake
shake quake flake baker rake flaky make shakily

Words with AME and AIM

The long **A** phoneme is heard in words with the letter string **AME**, for example, *fame, shameless* and *named*. When **ED** is added to verbs ending in **AME**, the **AME** letter string can still be seen, for example, *name – named*. When **ING** is added, the **E** is removed, for example, *tame – taming*.

The long **A** phoneme is also heard in the letter string **AIM**, for example, *aim, maim* and *claimant*, but this spelling is far less common.

The words in this crossword all contain the letter strings **AME** or **AIM**.

Word bank

FRAME	CAME
JAMES	DAME
AIMLESS	FAME
EXCLAIM	NAME
STAMENS	SAME
NAMESAKE	TAME
SHAMEFUL	BLAME
BLAMELESS	CLAIM
FRAMEWORK	FLAMES

ACROSS

2. To speak suddenly or cry out (7)
5. What you are called (4)
7. You see these in a fire (6)
10. A square of wood around a picture (5)
12. A person with the same name as someone else (8)
15. Having no direction or purpose (7)
16. A synonym for disgraceful (8)
17. Domesticated like a pet and not wild (4)
18. A supporting structure (9)

DOWN

1. A slang word for a woman and also a lady who has received an OBE award (4)
3. To call for, or demand as a right (5)
4. Innocent and not guilty (9)
6. The past tense of come (4)
8. The pollen producing parts of flowers (7)
9. Jim is a shortened form of this man's name (5)
11. Some people want '_ _ _ _ and fortune' (4)
13. Alike and not different (4)
14. To find fault with or accuse someone (5)

Words with AIN and ANE

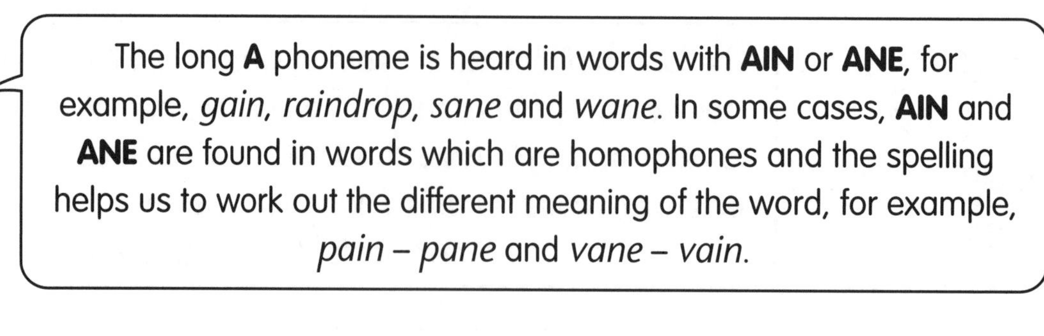

Using the word bank to help you, find an **AIN** or **ANE** word for each clue below, then circle or highlight the word in the wordsearch.

T	W	P	Q	E	N	T	E	R	T	A	I	N	E	D
C	O	M	P	L	A	I	N	Z	Y	S	M	J	S	J
P	A	I	N	T	E	K	T	R	R	T	R	L	V	H
S	A	M	T	N	E	T	K	F	P	A	Z	D	A	H
P	L	E	A	K	N	X	L	X	N	I	C	A	I	P
R	L	C	R	I	D	D	P	I	Z	N	E	N	N	A
A	R	D	A	O	N	C	A	L	L	N	P	E	Z	I
I	H	R	R	I	P	R	K	S	A	K	M	N	Y	N
N	T	R	A	A	G	L	F	J	P	I	I	T	F	F
N	R	R	R	R	I	N	A	E	R	A	N	K	A	U
C	B	T	N	A	L	N	N	N	R	E	I	L	I	L
G	M	A	N	E	I	A	R	T	E	R	M	N	N	X
V	N	T	M	Y	L	N	S	R	L	N	D	A	T	R
Y	C	H	A	I	N	C	O	N	T	A	I	N	I	V
H	U	R	R	I	C	A	N	E	D	E	T	A	I	N

Word bank

TRAIN	JANE
DETAIN	LANE
REMAIN	MANE
SPRAIN	RAIN
STRAIN	VAIN
CONTAIN	BRAIN
EXPLAIN	CHAIN
PAINFUL	DRAIN
COMPLAIN	FAINT
AEROPLANE	GRAIN
HURRICANE	PAINT
ENTERTAINED	SPAIN
CANE	STAIN
DANE	

1. A synonym for hurtful (7) **2.** A narrow road (4) **3.** To hold or have room for (7) **4.** To make a great effort or stretch tightly (6) **5.** A flying machine (9) **6.** A twisting injury that can hurt your ankle (6) **7.** A very strong and violent wind (9) **8.** The seeds of cereal plants, for example, oats and wheat (5) **9.** A type of transport that moves on rails (5) **10.** Someone from Denmark (4) **11.** A walking stick (4) **12.** This is protected by the skull (5) **13.** A series of connected rings often made of metal (5) **14.** To stay and not move away (6) **15.** To hold back or keep waiting (6) **16.** To talk about and make clear (7) **17.** A girl's name (4) **18.** To feel weak and collapse (5) **19.** The long hair on the neck of a lion or horse (4) **20.** To grumble or moan about something (8) **21.** A mark that cannot be removed (5) **22.** Conceited or big headed (4) **23.** To make a picture using brushes and oils or watercolours (5) **24.** A sewer or place where water flows out (5) **25.** Water that falls from the sky (4) **26.** The capital of this country is Madrid (5) **27.** The magician ____________ the children at the party (11)

Words with AY

The long **A** phoneme is sometimes spelt with the digraph **AY**, for example, *gateway*, *saying* and *playful*. This long **A** phoneme can also be spelt **EY** and **EIGH**, for example, *they* and *weigh*, but this spelling is far less common. The words *duvet* and *valet* also have long **A** phonemes, but they have completely different spellings, because they are French words.

The words in this puzzle all have a long **A** phoneme and contain the digraph **AY**. Read the clues and write the answers in the alien's spaceship. Use the word bank to help you.

1. A marine flatfish with fins like wings.
2. Giving money for goods or services.
3. A pet animal that is homeless or lost.
4. A race in which a baton is passed between runners.
5. To be disloyal or 'give the game away'.
6. Torn and ragged at the edges.
7. An underground passage or railway.
8. A flat board with a rim for carrying things.
9. Now, not yesterday or tomorrow.
10. A lazy person or loafer.

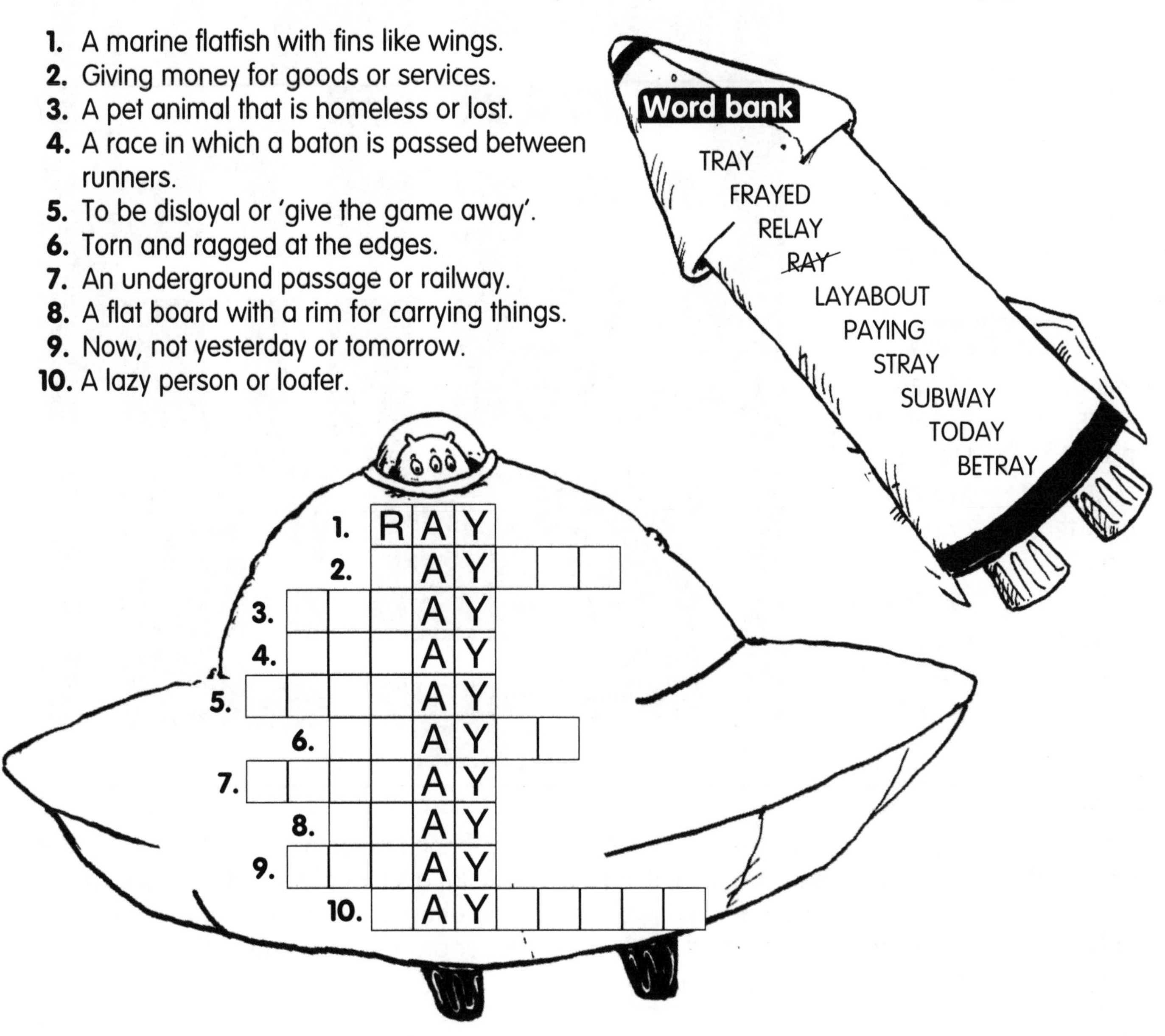

Find the ten **AY** words in the letter puzzle and write them in sentences on a separate piece of paper. The first one has been done for you.

aw ayst ayyes terday casta waybir thdaypay ment holid ayhay wireways is may

Words with ATE

The long **A** phoneme is heard in words with **ATE**, for example, *gateway, slate* and *exaggerate*. When **ED** is added to verbs ending in **ATE**, the letter string remains the same, for example, *concentr**ate** – concentr**ate**d*. When **ING** is added, the **E** is removed, for example, *date – dating*. The long **A** phoneme can also be heard in words with **EAT**, **AIT**, **EIGHT** and **AIGHT**, for example, *great, wait, weight* and *straight*.

The words in this crossword all have a long **A** phoneme and contain the letter string **ATE**.

Word bank

PRIMATE	ESTIMATE
IRATE	EXCAVATE
PLATE	IRRITATE
MATE	MOTIVATE
CREATE	CALCULATE
DEBATE	KATE
LATE	CANDIDATE
DONATE	CELEBRATE
SKATES	STATEMENT
ATE	PARTICPATE
MIGRATE	

ACROSS

4. To move to another place, as birds do (7) **6.** To make, bring into being (6) **7.** To encourage or prompt someone to do something (8) **9.** A flat dish for eating off (5) **10.** To annoy or inflame (8) **12.** To make a logical approximation of an amount (8) **13.** A friend or buddy (4) **17.** To compute or add up (9) **18.** To discuss or argue formally (6) **19.** To have festivities or a party for a special occasion (9)

DOWN

1. To join and take part (11) **2.** The past tense of eat (3) **3.** Not on time (4) **5.** To give money or food to charity (6) **6.** Someone standing for election (9) **8.** Boots with wheels or blades (6) **11.** A verbal or written report (9) **12.** To hollow out or dig a hole (8) **14.** A monkey or ape (7) **15.** A synonym for angry (5) **16.** A girl's name, short for Kathryn (4)

Words with EED, EAD and EDE

The long **E** phoneme is heard in words with the letter string **EED**, for example, *aniseed* and *freedom*. It can also be heard in words with **EAD** and **EDE**, for example, *lead* and *stampede*, but these letter strings are less common. The same long **E** phoneme can also be heard in the following words: *he'd, she'd* and *we'd*.

The words in this crossword all contain the letter string **EED** and a long **E** phoneme.

ACROSS

2. These are plants that gardeners like to get rid of (5) **7.** To be greater than (6) **8.** To go forward and carry on (7) **10.** Legal documents (5) **12.** Water plants with tall straight stems (5) **13.** To win and not fail (7) **15.** If you want too much money or food you are being ______ (6) **16.** Released or let go, for example, 'The prisoner was _____' (5) **18.** The present tense of fed (4) **19.** To want or require (4)

DOWN

1. Unnecessary (8) **3.** Fast (6) **4.** A deep cut will do this (5) **5.** Certainly, in fact (6) **6.** Consented or said yes (6) **9.** To take notice of or listen to (4) **11.** A plant that grows in the ocean and on rocks near the shore (7) **13.** You can sow these and plants will grow from them (5) **14.** To rear or generate (5) **17.** A rough strong cloth used for jackets and coats (5)

Words with ANGE

The letter string **ANGE** has a silent **E** and a soft **G**, for example, *strange, range* and *arrangement*. When **ED** is added to verbs ending in **ANGE**, the **ANGE** letter string remains the same, for example, *arrange – arranged*, but when **ING** is added, the **E** is removed, for example, *arrange – arranging*.

Find the missing **ANGE** words in the wordsearch and write them in the sentences below. Use the word bank to help you.

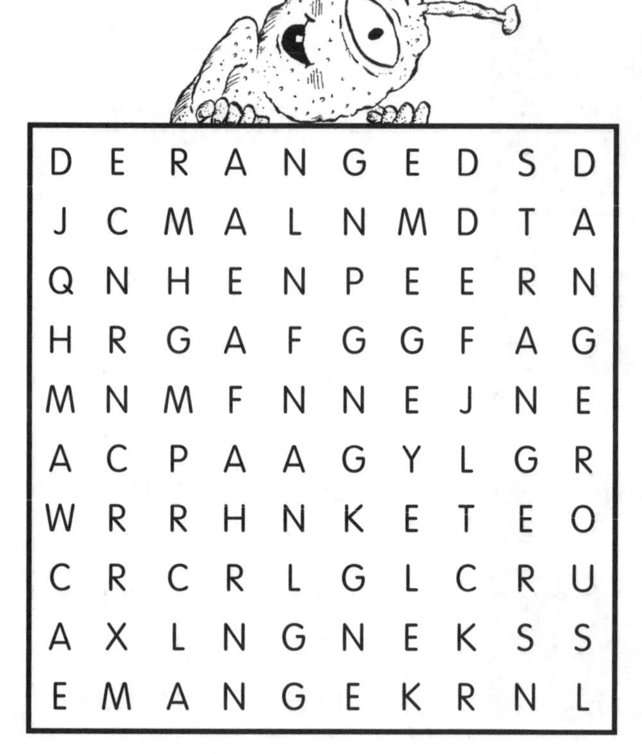

D	E	R	A	N	G	E	D	S	D
J	C	M	A	L	N	M	D	T	A
Q	N	H	E	N	P	E	E	R	N
H	R	G	A	F	G	G	F	A	G
M	N	M	F	N	N	E	J	N	E
A	C	P	A	A	G	Y	L	G	R
W	R	R	H	N	K	E	T	E	O
C	R	C	R	L	G	L	C	R	U
A	X	L	N	G	N	E	K	S	S
E	M	A	N	G	E	K	R	N	L

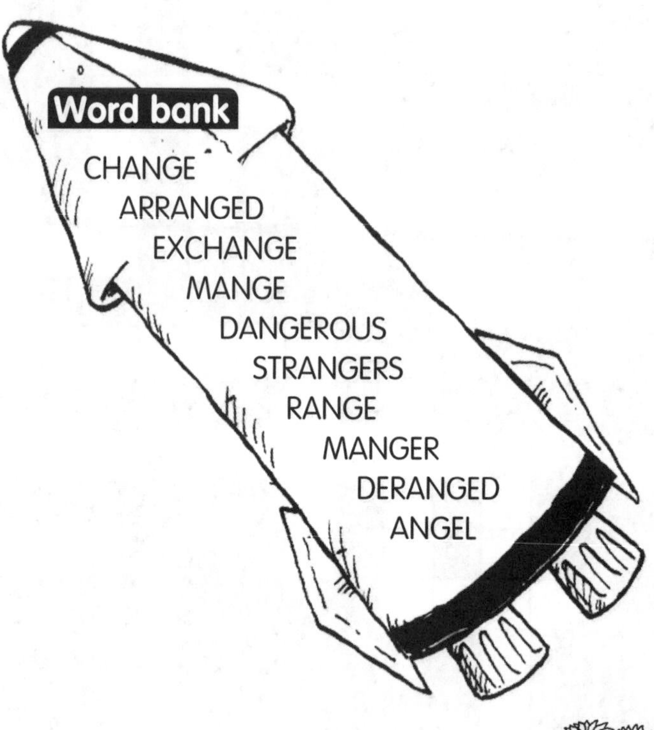

1. I _______________ the flowers in the vase and they looked beautiful. (8)

2. The old stray dog was covered in _______________ and his fur was falling out. (5)

3. In the Bible story, Jesus was placed in a _______________ in a stable. (6)

4. My mum always told me not to talk to _______________. (9)

5. Some people love to do extreme sports which can be very _______________. (9)

6. My mum said I was an _______________ when I brought her breakfast in bed. (5)

7. The mentally ill man was _______________ and did not know who he was or what he was doing in hospital. (8)

8. I like a wide _______________ of music, from classical to hip hop and rap. (5)

9. At the airport I was able to _______________ my foreign currency at a good rate. (8)

10. The actress had to do a quick _______________ behind the stage before she came on for the next act. (6)

Words with ANGE

Sort these **ANGE** words into groups in the table below. If they belong in more then one group, write them in each column.

arranged mange dangerously manger strangers
dangerous angel range exchange change strange grange
danger ranger strangely changeable mangy

Noun – singular	Noun – plural	Verb – present tense

Adverb	Adjective	Verb – past tense

Words with AIR and ARE

The letter string **AIR** or **ARE** is found in words such as, *hair, repair, glare* and *blared*. These letter strings also sound the same: **AIRE**, **ERE** and **EAR**, for example, *millionaire, there* and *bear*. In some cases, the different letter strings are found in words which are homophones and the spelling helps us to work out the difference between the two words, for example, *pear – pair – pare, hair – hare* and *bear – bare*.

Read the definition and choose the correct spelling, then write the word in the last column.

	Definition	Choice of spellings	Correct spelling
1	Two items.	pair, pare, pear	
2	Uncommon.	rare, rair, rear	
3	Something to sit on.	chare, chair, chear	
4	It grows on your head.	hare, hair, hear	
5	Naked.	bare, bair, bear	
6	To mend.	repare, repair, repear	
7	To frighten.	scare, scair, scear	
8	A bad dream.	nightmare, nightmair, nightmear	
9	A survey with questions.	questionnare, questionnaire, questionnair	
10	To put on the body.	ware, wair, wear	
11	Not here.	thare, thair, there	
12	A female horse.	mare, mair, mere	
13	A small imaginary creature with wings.	farey, fairy, feary	
14	Money paid to travel on a bus, train or plane.	fare, fair, fear	
15	A wild animal's den.	lare, lair, lear	
16	A bright light used as a signal.	flare, flair, flear	
17	To use crude or rude words.	sware, swair, swear	

Words with AIR and ARE

Choose the correct letter string (**AIR**, **ARE**, **EAR** or **ERE**) for these words and write them in the rockets below.

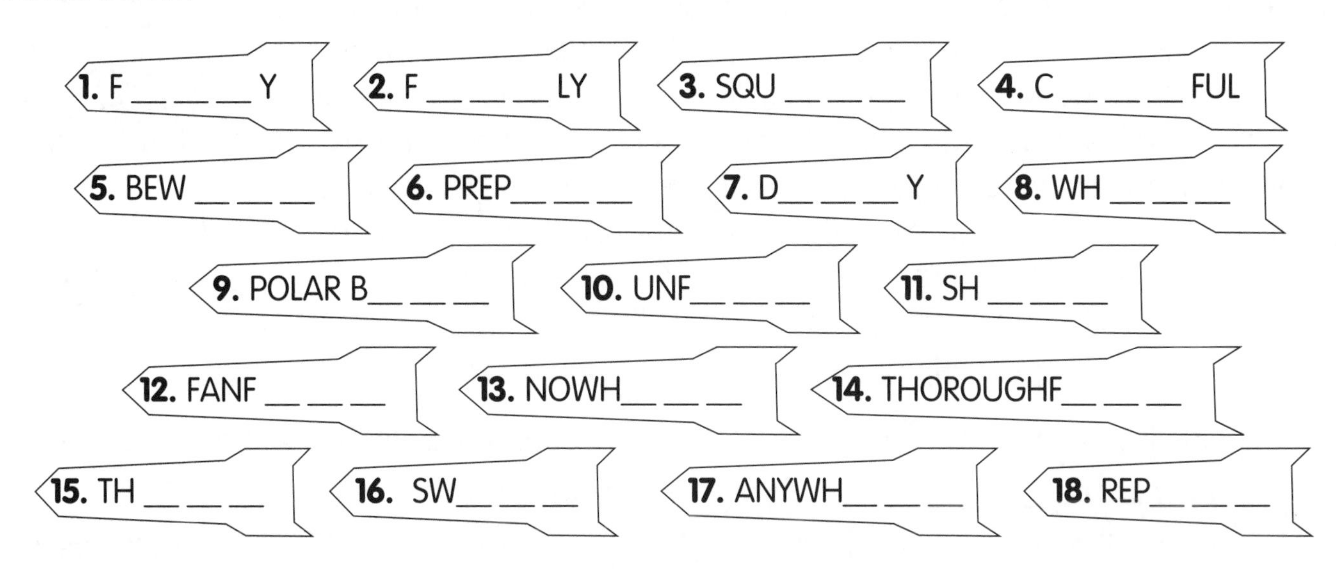

Write the missing words in these sentences. They all include the letter strings **AIRE**, **AIR**, **ARE**, **EAR** or **ERE**.

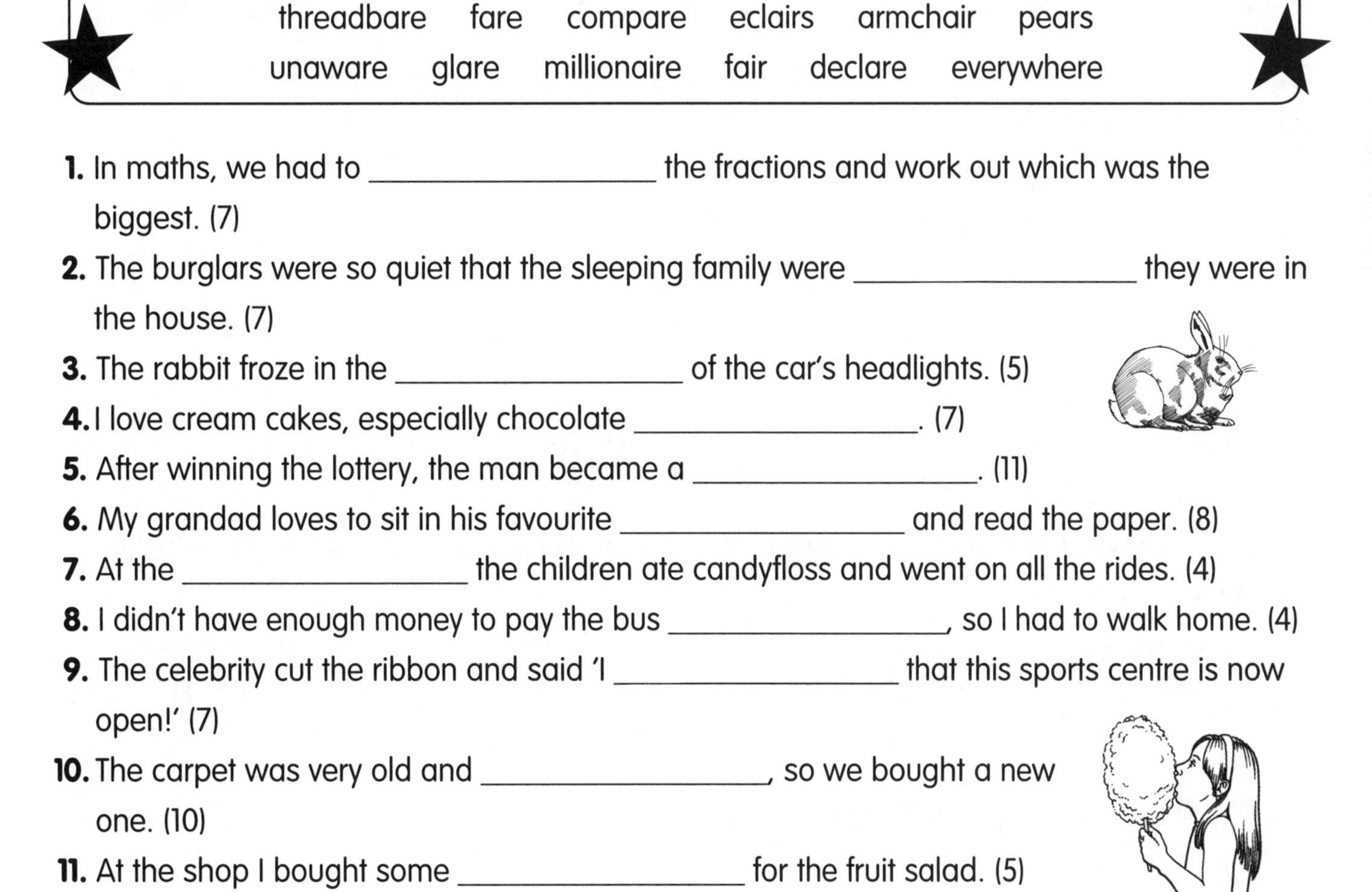

threadbare fare compare eclairs armchair pears
unaware glare millionaire fair declare everywhere

1. In maths, we had to ________________ the fractions and work out which was the biggest. (7)

2. The burglars were so quiet that the sleeping family were ________________ they were in the house. (7)

3. The rabbit froze in the ________________ of the car's headlights. (5)

4. I love cream cakes, especially chocolate ________________. (7)

5. After winning the lottery, the man became a ________________. (11)

6. My grandad loves to sit in his favourite ________________ and read the paper. (8)

7. At the ________________ the children ate candyfloss and went on all the rides. (4)

8. I didn't have enough money to pay the bus ________________, so I had to walk home. (4)

9. The celebrity cut the ribbon and said 'I ________________ that this sports centre is now open!' (7)

10. The carpet was very old and ________________, so we bought a new one. (10)

11. At the shop I bought some ________________ for the fruit salad. (5)

12. We looked ________________ for the missing money. (10)

Words with EAK and EEK

The long **E** phoneme is heard in words with **EAK** or **EEK**, for example, *beak* and *peek*. In some cases, the different letter strings are found in words which are homophones and the spelling helps us to work out the difference between the two words, for example, *peak – peek* and *weak – week*. Some words with **EAK** are unusual as they do not have a long **E** phoneme, for example, *steak* and *break*. This long **E** phoneme is also heard in the letter string **IQUE**, for example, *physique, unique* and *antique*, but this spelling is far less common.

Choose the correct spelling for the missing words in these sentences.

1. The old hinges began to ______________ as the door moved in the wind and I could hear the water in the ______________ flowing by. (**creak**, **creek**)
2. I climbed to the top of the mountain ______________ for a ______________ at the view. (**peak**, **peek**)
3. I needed lots of fresh ______________ to make my soup and I hoped the old pan didn't have any ______________. (**leaks**, **leeks**)
4. After I was sick I felt very ______________ for over a ______________. (**weak**, **week**)

Choose the correct spellings (**EAK** or **EEK**) for these words and put them into sentences on a separate piece of paper.

1. SQU_ _ _ **2.** FR_ _ _ **3.** M_ _ _

4. S_ _ _ **5.** SP_ _ _ **6.** STR_ _ _ **7.** WR_ _ _

8. BL_ _ _ **9.** GR_ _ _ **10.** CH_ _ _ **11.** SL_ _ _

Read the clues and add the suffixes **ING** or **Y** to the correct words below. Write the word on the line next to the clue.

SPEAK SQUEAK SEEK CHEEK

1. Noisy like a hinge that needs oiling. ______________
2. A synonym for talking. ______________
3. Impudent. ______________
4. Searching for something. ______________

Words with EAL and EEL

The long **E** phoneme is heard in words with **EAL** or **EEL**, for example, *heal, ideal, kneel* and *peel*. In some cases, the different letter strings are apparent in words which are homophones and the spelling helps us to work out the difference between the two words, for example, *heel – heal* and *reel – real*. This long **E** phoneme can also be heard in words with **ILE**, for example, *automobile* and *imbecile*, but this spelling is far less common.

Write the missing words in these sentences. They all include the letter strings **EAL** or **EEL**.

real	deal	meal	appeal	feel	conceal	steel	seal	keel	peel
reel	wheel	squeal	heal	peal	kneel	steal	veal	heel	ideal

1. I began to _______________ scared as the rollercoaster went higher and higher.
2. After her marathon run, she had a nasty blister on her _______________.
3. The boat's _______________ had a big hole in it and we began to sink.
4. We had to _______________ to say our prayers in church.
5. I wanted to _______________ my apple, as the skin was blemished.
6. For Christmas, my brother got a new fishing rod and _______________.
7. The saucepans were made of stainless _______________ and would not tarnish.
8. The front _______________ of my bike was badly dented in the accident.
9. I got a great _______________ when I bought my new computer and I paid much less than I expected.
10. For cuts to _______________ properly, they must be kept clean and dry.
11. My umbrella was _______________ for keeping the sun off my face.
12. My favourite _______________ is pizza and chips.
13. You could hear the _______________ of the church bells from miles away.
14. The antique wasn't _______________; it was a fake.
15. To make my baby cousin _______________, all I have to do is tickle her.
16. My mum told me it is very wrong to _______________ things from other people.
17. The meat from young calves is called _______________.
18. I had to _______________ to my mum to let me go to the party.
19. The baby _______________ had lost its mother and got washed up on the beach.
20. I used some touch-up paint to _______________ the scratch I had made on my dad's car.

Words with EAP and EEP

The long **E** phoneme is heard in words with **EEP**, for example, *steep* and *sleep*. It can also be heard in the letter string **EAP**, for example, *heap* and *leap*, but this spelling is much less common. *Weapon* is an exception with a short **E** phoneme.

Look at these pairs of words and write the correct spelling in the empty column.

	Choice of spellings		Correct spelling
1	DEEPER	DEAPER	
2	KEEP	KEAP	
3	SLEAP	SLEEP	
4	STEEPLY	STEAPLY	
5	HOUSEKEAPER	HOUSEKEEPER	
6	CREAPER	CREEPER	
7	REEP	REAP	
8	SQUEEL	SQUEAL	

Sort these words into the column below. If they belong in more than one group, write them in more than one column.

deeper deep heaps beep leap sleep sweeper
sheep deepest deeply steeply creeper heap housekeeper
heaped beeped weep sleep peep cheaply cheapest
peeped asleep steep jeep keep sheep

Noun – singular	Noun – plural	Verb – present tense	Verb – past tense	Adjective	Adverb

Words with EER and EAR

The long **E** phoneme is heard in words with **EER** or **EAR**, for example, *cheer, deer, clear* and *fear*. It can also be heard in the letter strings **ERE**, **EIR** and **IER**, for example, *here, weir* and *pier*. **EER** is also a suffix which turns a word into the person doing an action, for example, *mountaineer* and *volunteer*.

The words in this crossword have the letter strings **EER**, **EAR** and **ERE** and the long **E** phoneme.

Word bank

CHEERFUL	CLEAR
SMEAR	STEER
ENGINEER	FEAR
MOUNTAINEER	PEER
VOLUNTEER	MERE
REINDEER	HEAR
SNEER	SPEAR
MUSKETEERS	YEAR
GEARS	JEER
DREARY	ERE
DISAPPEAR	DEAR
TEARFUL	

ACROSS

2. A long pointed weapon (5) **6.** Before (of time) (3) **8.** You use your ears to do this (4) **9.** A scornful or contemptuous expression (5) **10.** 12 months (4) **12.** To look closely (4) **13.** Sir Edmund Hillary became a very famous one when he climbed Everest (11) **14.** To guide or direct the course of a vehicle (5) **16.** Dull and dismal (6) **20.** Precious or expensive (4) **21.** Happy and contented (8) **22.** Someone who offers their service without being asked and often without payment (9) **23.** To vanish out of sight (9)

DOWN

1. In the legend there were three of these men with guns (10) **3.** The male of this species has antlers (8) **4.** To mock scoff or taunt (4) **5.** Insignificant or minor, for example, her injury was a _ _ _ _ scratch (4) **7.** A person who is in charge of engines and machinery (8) **11.** To be afraid is to have _ _ _ _ (4) **15.** To be upset (7) **17.** Cars can have four or more of these which can be changed manually or automatically (5) **18.** Transparent or obvious (5) **19.** To smudge or rub with grease (5)

Words with EAT and EET

The long **E** phoneme is heard in words with **EAT**, for example, *cheat* and *meat*. It can also be heard in the letter string **EET**, for example, *sweet* and *feet*. This long **E** phoneme is also found in words containing **ETE** (with a silent **E**), for example, *concrete* and *delete*, but this spelling is far less common. The words *great* and *threat* are exceptions as they have the **EAT** letter string but do not have a long **E** phoneme.

Look at these pairs of words and write the correct spelling in the empty column.

	Choice of spellings		Correct spelling
1	CHEET	CHEAT	
2	CHEETAH	CHEATAH	
3	WHEET	WHEAT	
4	REPEET	REPEAT	
5	DEFEET	DEFEAT	
6	HEET	HEAT	
7	SHEET	SHEAT	
8	SLEET	SLEAT	
9	MEETING	MEATING	
10	SEAT	SEET	

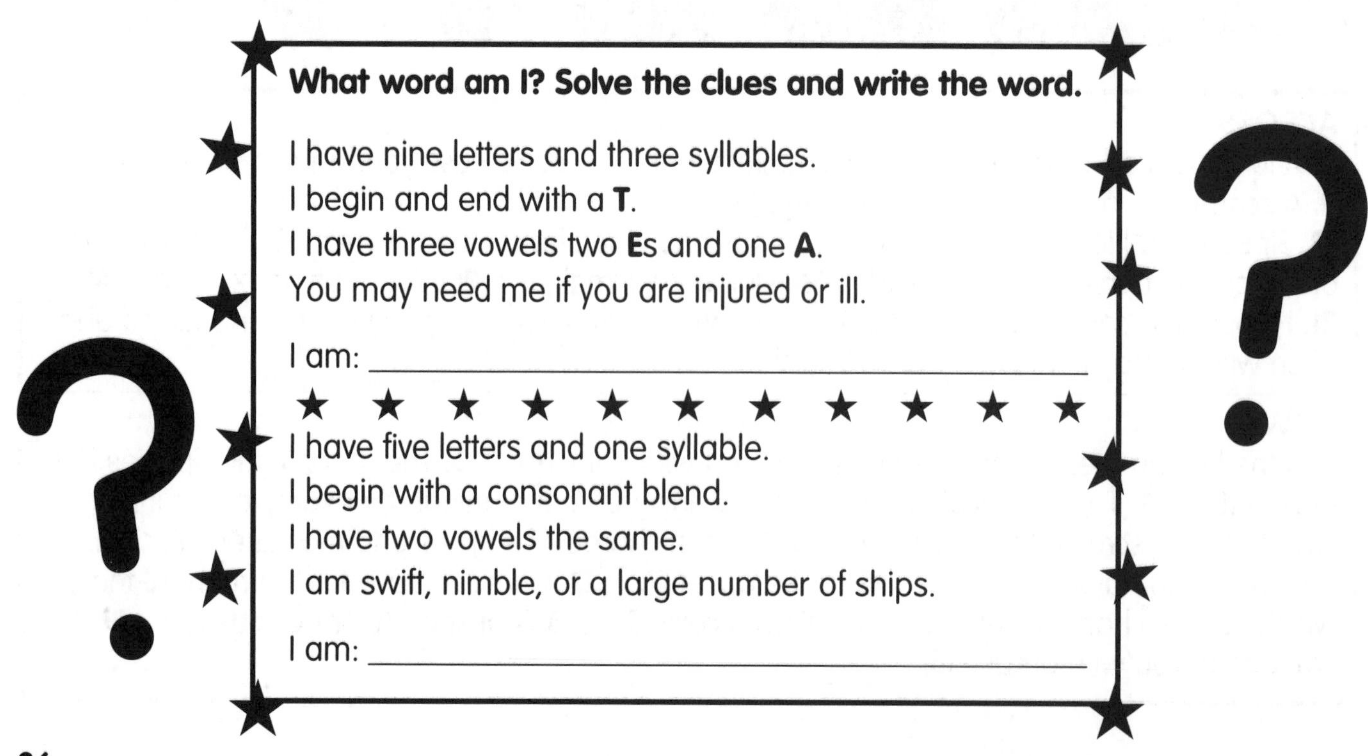

What word am I? Solve the clues and write the word.

I have nine letters and three syllables.
I begin and end with a **T**.
I have three vowels two **E**s and one **A**.
You may need me if you are injured or ill.

I am: ______________________________

I have five letters and one syllable.
I begin with a consonant blend.
I have two vowels the same.
I am swift, nimble, or a large number of ships.

I am: ______________________________

Words with **EAT** and **EET**

Solve the clues below and then highlight the answers in the wordsearch. All the answers contain **EAT** or **EET**. Use the word bank to help you.

K	M	B	E	A	T	E	N	Z	Q	Q	Z	X	P	W	R
R	S	L	B	H	T	R	E	P	E	A	T	Z	H	Z	J
K	T	H	G	E	R	W	T	Z	X	H	T	L	E	N	L
M	Z	P	E	T	E	N	W	R	Z	D	T	R	A	P	Q
R	M	R	R	E	A	T	G	R	E	E	T	T	R	L	M
R	X	F	T	K	T	V	L	Y	P	A	A	R	T	E	V
K	V	W	P	K	Y	S	D	E	R	E	T	X	B	A	R
S	E	A	T	B	E	L	T	F	N	C	T	Y	E	T	P
K	C	M	C	S	E	A	T	Z	Y	A	D	G	A	E	P
L	H	E	D	Y	J	O	R	L	E	N	E	K	T	D	N
D	E	A	L	J	O	Y	R	H	K	B	F	M	Z	T	V
Y	A	T	L	R	N	E	W	B	F	E	E	T	H	R	T
H	T	H	T	E	T	H	M	L	L	Q	A	L	D	A	B
D	E	E	T	A	P	N	K	B	C	E	T	T	E	T	L
J	E	A	E	R	K	K	X	R	M	T	A	B	M	P	L
B	E	H	T	C	M	K	N	L	W	J	N	T	H	R	C

1. To do or say something twice (6) **2.** A special pleasure (5) **3.** Tidy and orderly (4) **4.** Chewed and swallowed (5) **5.** Large pieces of material used on beds (6) **6.** This should be worn in a car to prevent injury in a crash (8) **7.** To say hi or hello (5) **8.** The rhythm in music or to do better than someone in a game (4) **9.** A signed contract between people or countries (6) **10.** An antonym for victory (6) **11.** Material that has lots of little folds is _ _ _ _ _ _ _ (7) **12.** An object for sitting on (4) **13.** This can be ground to make flour (5) **14.** You can count your pulse to measure this (9) **15.** The plural of foot (4) **16.** To play unfairly and break the rules of a game (5) **17.** Warmth or an elimination race (4) **18.** A dark red root vegetable eaten with salad (8) **19.** A machine used to warm things (6) **20.** An insect with hard upper wings (6) **21.** Whipped (as eggs can be) or hit repeatedly (6) **22.** Pork, beef and lamb are types of this (4) **23.** The cry of a lamb (5)

Words with ICE

The long **I** phoneme is heard in words with **ICE**, for example, *nicely*, *spice* and *priced*. When **ED** is added to verbs ending in **ICE**, the **ICE** letter string remains the same, for example, *price – priced*. When **ING** and **Y** are added, the **E** is removed, for example, *price – pricing* and *spice – spicy*. When **ICE** is at the end of words with more than one syllable, the **I** phoneme sometimes becomes short, for example, *practice* and *avarice*.

The words in this crossword all contain the letter string **ICE**.

Word bank

DICE	TRICE
LICE	TWICE
ICE CREAM	ADVICE
MICE	BODICE
NICE	DEVICE
RICE	ENTICE
VICE	MALICE
ALICE	NOTICE
PRICE	THRICE
SLICE	CHALICE

ACROSS

1. A gadget or tool (6) **4.** Information or a recommendation given by one person to another (6) **5.** Hatred or bad feeling (6) **8.** The girl who was in Wonderland (5) **10.** The upper part of a dress (6) **11.** The plural of louse (4) **13.** The cost of something (5) **16.** Two times (5) **17.** Three times (6) **18.** The plural of mouse (4) **19.** To attract or tempt (6)

DOWN

2. A poetic word for a cup (7) **3.** Cubes with dots on for playing board games (4) **6.** A device for holding things while working on them (4) **7.** A thin flat piece of something (5) **9.** A frozen dairy dessert (8) **12.** A sign or piece of paper with information on (6) **14.** This cereal is used to make risotto (4) **15.** Pleasant and not nasty (4) **16.** Very quickly 'In a _ _ _ _ _' (5)

Using the answers in the crossword, sort the words into two groups: 'Long **I** phoneme' and 'Short **I** phoneme'.

Words with IDE and IED

The long **I** phoneme is heard in the letter string **IDE** (with a silent **E**), for example, *side* and *pride*. It can also be heard in words with **IED** in the past tense, for example, *deny – denied* and *spy – spied.*

Find an **IDE** or **IED** word for each clue below, then circle or highlight your answers in the wordsearch.

Word bank

HIDE	SLIDE
DIED	TRIED
RIDE	DECIDE
TIDE	DIVIDE
TIED	INSIDE
WIDE	RESIDE
BRIDE	CONFIDE
CHIDE	DIOXIDE
CRIED	OUTSIDE
FRIED	PROVIDE
OXIDE	SUBSIDE
PRIDE	

1. To share or split up into parts (6) **2.** To tell a secret to someone (7) **3.** The rise and fall of the sea (4) **4.** To decrease or get less (7) **5.** We breathe out carbon _ _ _ _ _ _ _ (7) **6.** To choose or commit yourself to one option (6) **7.** The past tense of try (5) **8.** A compound of oxygen and one other element (5) **9.** An antonym for inside (7) **10.** They say it comes before a fall (5) **11.** The past tense of tie (4) **12.** To cater for or supply (7) **13.** To keep out of sight and also the name for the skin of an animal (4) **14.** A woman who is getting married (5) **15.** To live somewhere (6) **16.** An antonym for narrow (4) **17.** An antonym for outside (6) **18.** The past tense of cry (5) **19.** You do this on a horse (4) **20.** The past tense of fry (5) **21.** To scold or tell off (5) **22.** An antonym for lived (4) **23.** The slippery part of a children's playground (5)

Find the **IDE** and **IED** words in this letter puzzle and write a sentence for each one on a separate piece of paper. The first one has been done for you.

st ride c olli desatis fiedgu ided riedl ieds uppli eddieds pied

Words with IGHT and ITE

The long **I** phoneme is heard in words with **IGHT** or **ITE**, for example, *right* and *write*. When **ED** is added to verbs ending in **ITE**, the letter string remains unchanged, for example, *invite – invited*. When **ING** is added, the **E** is removed, for example, *invite – inviting*. Verbs with the letter string **ITE** are often irregular in the past tense, for example, *bite – bit* and *write – wrote*.

The words in this crossword all contain the letter strings **IGHT** and **ITE** and have a long **I** phoneme.

Word bank

BITE	TIGHT
POLITE	FRIGHT
WHITE	KNIGHT
WRITE	MIGHTY
STALACTITE	SLIGHT
KITE	DELIGHT
UNITE	EYESIGHT
HIGH	MIDNIGHT
FIGHT	TWILIGHT
LIGHT	LIGHTNING
NIGHT	HEADLIGHTS
RIGHT	

ACROSS

4. An antonym for left (5) **8.** To puts words on paper (5) **9.** The time after sunset before night (8) **11.** Not loose (5) **12.** Strong and powerful 'High and _ _ _ _ _ _' (6) **17.** Some people need glasses to improve this (8) **19.** An antonym for rude and ill-mannered (6) **20.** A soldier in armour and a chess piece (6) **21.** A shock or sudden fear (6) **22.** Not dark or heavy (5) **24.** An antonym for day (5)

DOWN

1. You use your teeth to do this (4) **2.** A stony spike hanging from the roof of a cave (10) **3.** Boxers do this (5) **5.** A toy flown in the wind at the end of a long piece of string (4) **6.** The colour of snow (5) **7.** The opposite of midday (8) **10.** An antonym for low (4) **13.** A car has two of these to put on in the dark (10) **14.** Small or slim (6) **15.** This often goes with thunder (9) **16.** To please greatly (7) **18.** To join together (5)

Words with IGHT and ITE

Say each word out loud and then write it in the table below.

quite	requisite	definite	bright
opposite	tonight	excite	infinite
midnight	invite	exquisite	granite

Long I phoneme		Short I phoneme	

Join a syllable from each column to make two-syllable words with **ITE** and **IGHT** and write them in the last column. One has been done for you.

1st syllable	2nd syllable	Complete word
spot	spite	
de	sight	
in	light	spotlight
air	cite	
hind	vite	
re	tight	

Join a syllable from each column to make three-syllable words with **ITE** and **IGHT** and write them in the last column. One has been done for you.

1st syllable	2nd syllable	3rd syllable	Complete word
ap	pe	ful	
co	en	ing	
fright	light	right	
de	py	tite	appetite

Words with ILE and IAL

The long **I** phoneme is heard in words with **ILE**, for example, *pile* and *smile*. It can also be heard in the letter string **IAL**, for example, *trial* and *dial*. When **ED** is added to verbs ending in **ILE**, the **ILE** letter string is unchanged, for example, *smile* – *smiled*. When **ING** is added, the **E** is removed, for example, *smile – smiling*. The letter string **ILE** can sometimes have a long **E** phoneme, for example, *automobile* and *imbecile*.

Look at these pairs of words with **ILE** and **IAL** and write the correct spelling in the empty column.

	Choice of spellings		Correct spelling
1	TRILE	TRIAL	
2	FILE	FIAL	
3	SMILE	SMIAL	
4	SUNDILE	SUNDIAL	
5	WHILE	WHIAL	
6	STILE	STIAL	
7	EXILE	EXIAL	
8	RECONCILE	RECONCIAL	
9	CAMOMILE	CAMOMIAL	
10	DENILE	DENIAL	

Unjumble these **ILE** and **IAL** anagrams and write the word next to it. Then write a short sentence for each one on a separate piece of paper.

1. N G I L A D L I ____________
2. C I O C O E D R L ____________
3. R I O E L F P ____________
4. L E I T R A D L ____________
5. E I N A L D ____________
6. D E I S L M ____________

Words with IRE

The long **I** phoneme is heard in words with **IRE** (with a silent **E**), for example, *fire* and *umpire*. Lots of words end in **SPIRE**, for example, *aspire, inspire, conspire, transpire* and *perspire*. 'Spire' comes from the Latin word 'spirare' which means 'to breathe'. When **ED** is added to verbs ending in **IRE**, the **IRE** letter string can still be seen, for example, *hire – hired*. When **ING** is added, the **E** is removed, for example, *hire – hiring*.

The words in this crossword all contain the letter string **IRE** and have a long **I** phoneme.

Word bank

FIRE
HIRE
INSPIRED
SPIRE
TIRE
UMPIRE
VAMPIRE
ADMIRE
ATTIRE
WIRE
BONFIRE
CONSPIRE
PERSPIRE
SAPPHIRE
WILDFIRE
INQUIRE
REQUIRE
ACQUIRE

ACROSS

2. Clothing(6) **5.** A bright blue precious stone (8) **6.** A match will help you start one of these (4) **8.** To ask or seek information (7) **10.** To need or want (7) **12.** To pay for the temporary use of something or someone (4) **14.** If something moves fast and out of control we say 'It spreads like _________' (8) **15.** To sweat (8) **17.** To look up to (6)

DOWN

1. A large outdoor fire (7) **2.** To obtain (7) **3.** Metal made into thin flexible strands (4) **4.** A type of bat that sucks the blood of animals (7) **7.** To get weary or be lacking in energy (4) **9.** To be encouraged, heartened or stimulated (8) **11.** A referee or the person who enforces the rules in a game (6) **13.** To collude or plot together (8) **16.** The pointed part of a steeple (5)

Words with INE

The letter string **INE** is very versatile as it can be pronounced in different ways:

- Long **I** phoneme, for example, *line* and *diner*.
- Short **I** phoneme, for example, *determine, discipline* and *masculine*.
- Long **E** phoneme, for example, *tambourine, pristine* and *Philippines*.

When **ED** is added to verbs ending in **INE**, the letter string remains the same, for example, *wh**ine** – wh**ine**d*. When **ING** is added, the **E** is removed, for example, *whine – whining*.

Find the ten **INE** words in the letter puzzle and write them on the lines below. They all have a long **I** phoneme. The first one has been done for you. On a separate piece of paper, write a sentence for each word or put more than one word in each sentence.

inc line o utlin enine teencan ines pinew inepi nepin e apple ref inel in e

____

____

Sort these words into groups in the table below. They all contain **INE** but this letter string is pronounced in different ways. Read each word aloud first.

airline determine combine decline discipline porcupine
submarine spine medicine heroine machine nectarine
ultramarine recline tambourine engine adrenaline incline
chlorine headline gasoline feminine margarine imagine

Long I phoneme	Short I phoneme	Long E phoneme

Words with INE

The words in this puzzle all have a long **I** phoneme and contain the letter string **INE**. Solve the clues and write the words in the spaceship. Use the word bank to help you.

1. My teacher told me to ______________ the date neatly with a ruler. (9)
2. A sacred place. (6)
3. A pH scale shows levels of acid and ______________. (8)
4. An area of land or sea containing explosives. (9)
5. Cats are feline animals and dogs are ______________. (6)
6. Cowardly, lily-livered or faint hearted. (9)
7. A type of cord string or yarn. (5)
8. 9 x 10 = ______________. (6)
9. A deep hole where a precious metal is dug out. (8)
10. Salty water. (5)

SHRINE MINEFIELD CANINE UNDERLINE BRINE
GOLDMINE NINETY ALKALINE TWINE SPINELESS

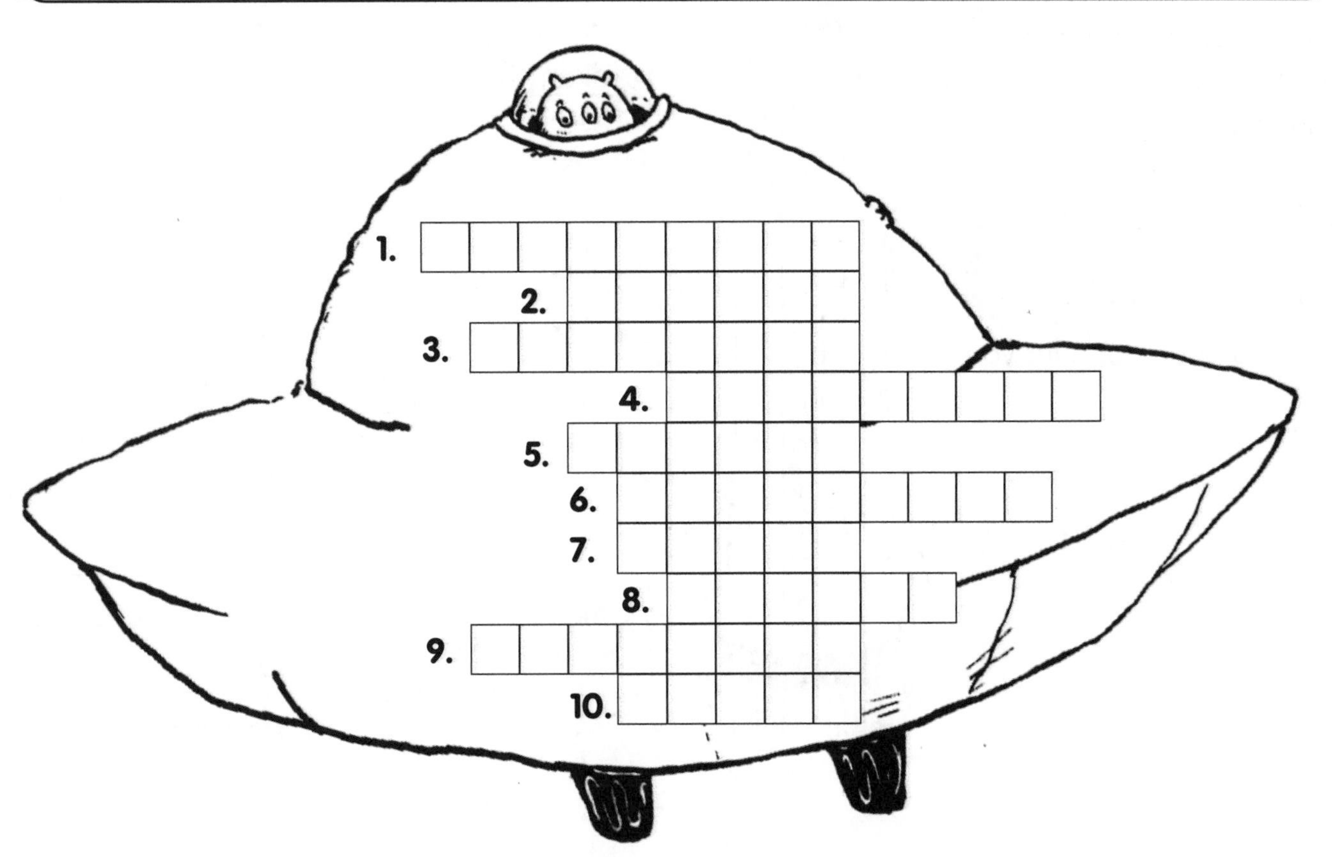

Words with OKE

The long **O** phoneme is heard in words with **OKE**, for example, *broke, sunstroke* and *smoke*. It is also heard in the letter strings **OAK** and **OLK**, for example, cloak, *soak, oak, yolk* and *folk*. When **ED** is added to verbs ending in **OKE**, the letter string remains the same, for example, ***joke*** – *jok**ed***. When **ING** is added, the **E** is removed, for example, *joke – joking*.

Where is the **OKE**? Look at these words and work out which syllable contains the **OKE** letter string. Sort them into groups in the table below. Two words have been written in for you.

~~CHOKER~~ BACKSTROKE SPOKEN PROVOKE STROKE ~~SUNSTROKE~~ BROKEN
STOCKBROKER YOKEL POKER EVOKE OUTSPOKEN PAWNBROKER AWOKEN

1st syllable OKE		2nd syllable OKE	
CHOKER		SUNSTROKE	

Choose letters from the rocket to join to a letter string in the planets to make a word. There are:

- 3 words ending in **OKEN**;
- 4 words ending in **VOKE**;
- 5 words ending in **OKER**.

One has been done for you.

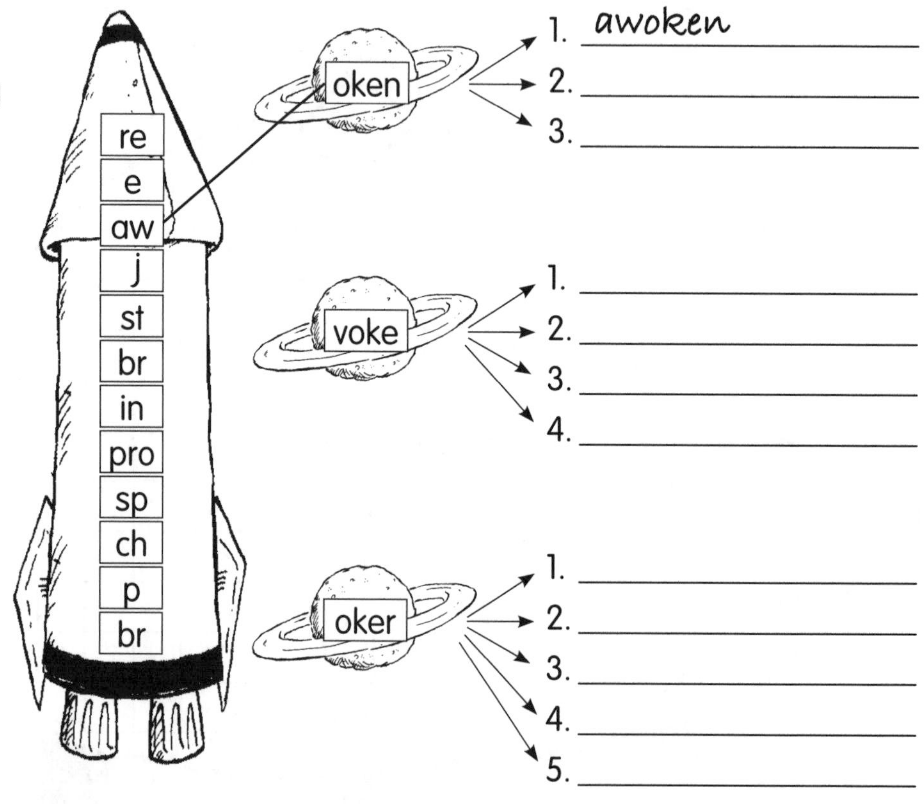

Words with **OKE**

Find an **OKE** word for each clue below. Then circle or highlight the words in the wordsearch.

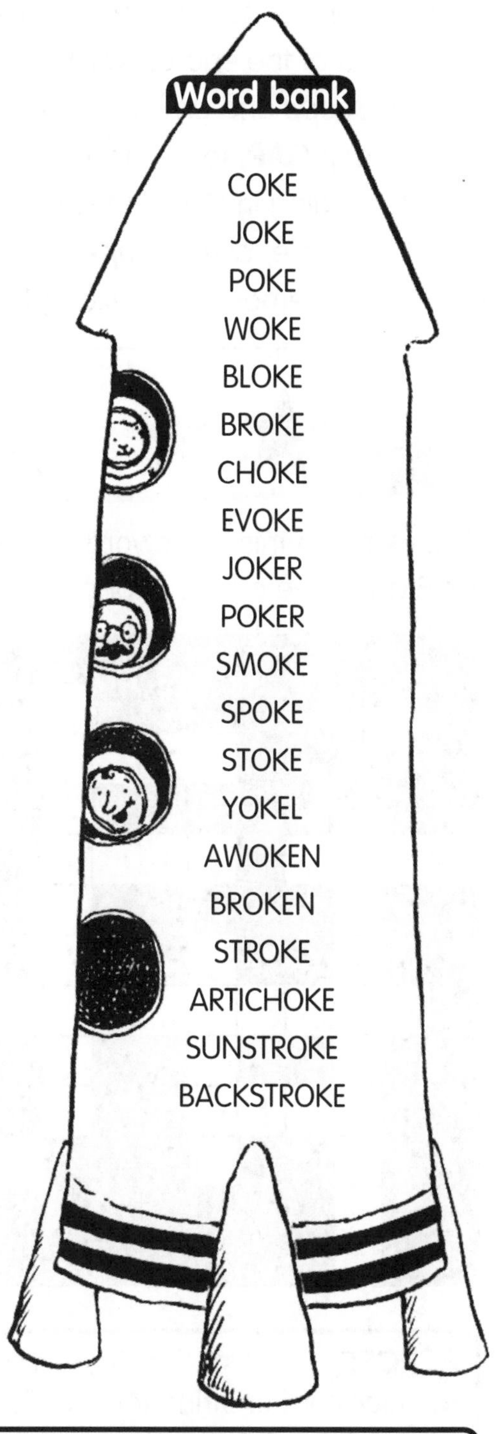

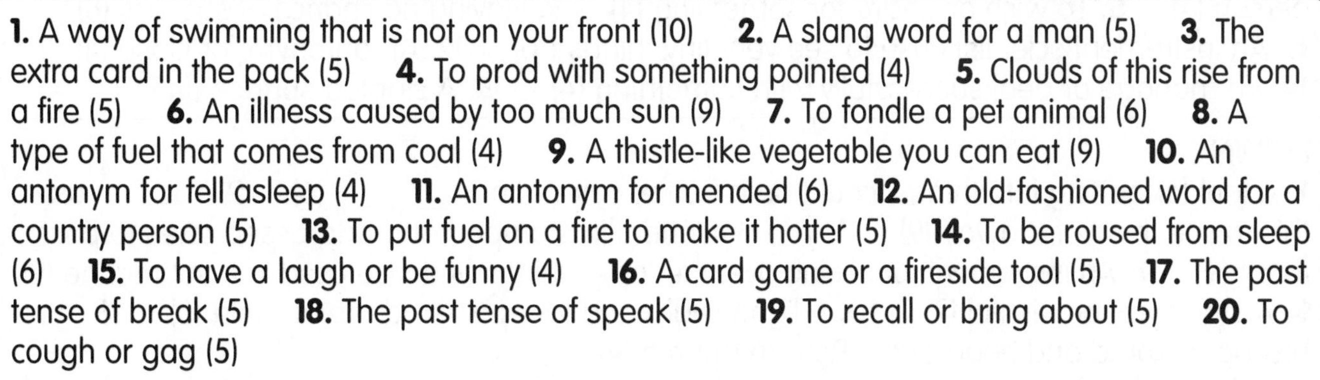

1. A way of swimming that is not on your front (10) **2.** A slang word for a man (5) **3.** The extra card in the pack (5) **4.** To prod with something pointed (4) **5.** Clouds of this rise from a fire (5) **6.** An illness caused by too much sun (9) **7.** To fondle a pet animal (6) **8.** A type of fuel that comes from coal (4) **9.** A thistle-like vegetable you can eat (9) **10.** An antonym for fell asleep (4) **11.** An antonym for mended (6) **12.** An old-fashioned word for a country person (5) **13.** To put fuel on a fire to make it hotter (5) **14.** To be roused from sleep (6) **15.** To have a laugh or be funny (4) **16.** A card game or a fireside tool (5) **17.** The past tense of break (5) **18.** The past tense of speak (5) **19.** To recall or bring about (5) **20.** To cough or gag (5)

Words with OPE

The long **O** phoneme is heard in words with **OPE**, for example, *coped*, *stethoscope* and *hoped*. **OPE** is often found at the end of words, for example, *microscope*, *telescope* and *envelope*. The phoneme **OPE** can also be spelt **OAP**, for example, *soap*. When **ED** is added to verbs ending in **OPE**, the letter string remains unchanged, for example, *cope – coped*. When **ING** is added, the **E** is removed, for example, *cope – coping*.

The words in this crossword all contain the letter string **OPE** and have a long **O** phoneme.

Word bank

COPE
DOPE
HOPE
MOPE
OPEN
POPE
ROPE
MOPED
SLOPE
OPENER
HOPEFUL
ENVELOPES
HOPELESS
PERISCOPE
MICROSCOPE
TELESCOPE

ACROSS

4. Thick cord or string (4) **5.** To be gloomy and apathetic (4) **6.** A slang word for a stupid person (4) **8.** To wish or desire for something (4) **9.** Having no chance of success (8) **11.** An instrument scientists use to see very tiny things (10) **12.** An antonym for close (4) **14.** To manage or deal successfully with something (4) **15.** A slanted surface (5)

DOWN

1. A tool for getting into things, could be a bottle, can or tin _ _ _ _ _ _ (6) **2.** You can use this to see things far away (9) **3.** You put letters in stamped and addressed ones to post them (9) **7.** An instrument for seeing over the top of things and also used in a submarine (9) **9.** A synonym for wishful (7) **10.** A lightweight and low-powered motor cycle (5) **13.** A bishop of Rome and head of the Roman Church (4)

Words with OTE and OAT

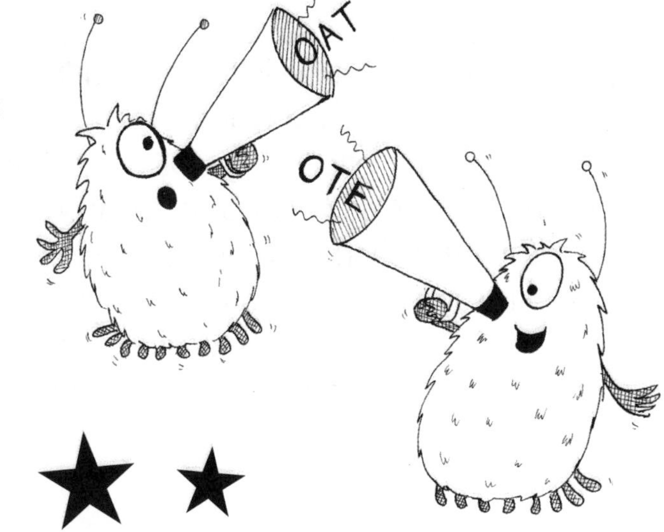

The long **O** phoneme is heard in the letter strings **OTE** or **OAT**, for example, *note, vote, goat* and *boat*. When **ED** is added to verbs ending in **OTE**, the letter string remains unchanged, for example, *vote – voted*. When **ING** is added, the **E** is removed, for example, *vote – voting*.

Put these missing words in the sentences below.

anecdote	wrote	afloat	demote	goat	petticoats	coat	throat
lifeboat	notes	remote	bloated	gloat	devoted	votes	quote

1. It was hard to stay ______________ in the rough sea; I was so glad to see the ______________ coming to our rescue.

2. The doctor looked down my ______________ when I lost my voice and ______________ lots of ______________ on his computer.

3. An ______________ is a short account of an incident.

4. When I went to the boss's office I was afraid he was going to ______________ me, but luckily I got promoted instead.

5. I liked the country cottage; I didn't even mind the pet ______________ but it was too ______________ for me as I wanted to be closer to town.

6. In the olden days, women used to wear ______________ under their skirts.

7. After the huge five course meal I felt really ______________ and I couldn't even do up my ______________.

8. The boy was ______________ to his mother.

9. In my English Literacy exam I tried to ______________ some famous authors.

10. When I won the election with nearly all the ______________, it was hard not to ______________ and get big-headed.

Find the **OAT** and **OTE** words in this letter puzzle and write them in sentences on a separate piece of paper. The first one has been done for you.

prom otevo tem otorbo atgloa tanti dotede voter otegloat ingnoted

Words with OOD and UDE

The phoneme in the letter string **OOD** can sound short, for example, *good* and *wood*, or long, for example, *food* and *mood*. The letter string **UDE** has a long **O** phoneme, for example, *rude* and *conclude*. The word *blood* is an exception, as it has a short **U** phoneme. It should also be noted, that **OO** is sometimes pronounced in different ways according to regional dialects.

Find and circle the words in this letter puzzle and write them in the spaceship. (The words below all include the phoneme **OO** but they are not in the same order as the words in the spaceship.)

MIS UNDERS TOODWO ODENHO ODWINK CHILD HO ODMO ODY NEI GHBO UR HOODUN DERSTO ODWIT HSTO ODWO ODGO OD

Words with OOD and UDE

Sort these words according to their **OO** phoneme and write them in the table.

neighbourhood good childhood livelihood wood hoodwink
stood misunderstood mood broody food withstood

Short OO phoneme		Long OO phoneme

The words in this table use **OOD** or **UDE**. Choose the correct spelling for each word and write it in the empty column.

	Choice of spellings		Correct spelling
1	CRUDE	CROOD	
2	LIKELIHUDE	LIKELIHOOD	
3	MOTHERHUDE	MOTHERHOOD	
4	INTRUDER	INTROODER	
5	SECLUDED	SECLOODED	
6	INTERLUDE	INTERLOOD	
7	KNIGHTHUDE	KNIGHTHOOD	
8	GUDENESS	GOODNESS	
9	UNDERSTUDE	UNDERSTOOD	
10	EXCLUDED	EXCLOODED	

Words with OON

The letter string **OON** is found in words such as *afternoon, balloon* and *spoon*. These letter strings also sound the same: **UNE** and **EWN**, for example, *June* and *strewn*. The letter string **UNE** is usually pronounced with a long **U** phoneme rather than a long **OO** phoneme. Say these words aloud to hear the difference: *immune, moon, tune* and *platoon*.

Sort these **OON** and **UNE** words according to the number of syllables they have and write them in the table below.

DESSERTSPOON MONSOON SWOON SCHOONER PRUNE
BALLOONS MISFORTUNE HONEYMOON TYPHOON SPOON DUNE
CROON HARPOONS FORTUNE TEASPOON MOON PLATOON
TABLESPOON AFTERNOON LAGOON CARTOONS TUNE SOON
NOON MACAROON IMMUNE

1 syllable	2 syllables	3 syllables

Words with OON

Using the word bank to help you, complete the sentences below and then find the words in the wordsearch. All the words contain the letter strings **OON** or **UNE**.

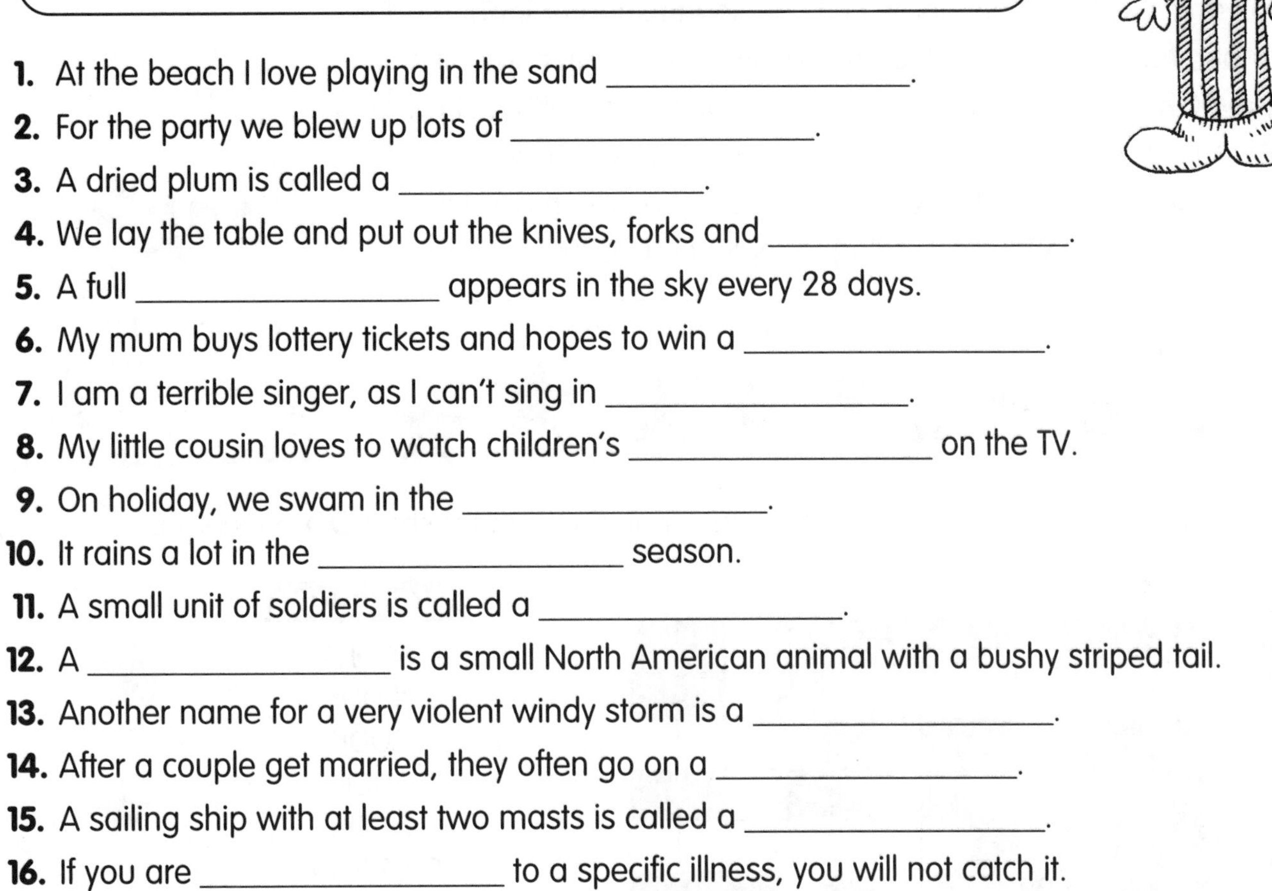

1. At the beach I love playing in the sand ________________.
2. For the party we blew up lots of ________________.
3. A dried plum is called a ________________.
4. We lay the table and put out the knives, forks and ________________.
5. A full ________________ appears in the sky every 28 days.
6. My mum buys lottery tickets and hopes to win a ________________.
7. I am a terrible singer, as I can't sing in ________________.
8. My little cousin loves to watch children's ________________ on the TV.
9. On holiday, we swam in the ________________.
10. It rains a lot in the ________________ season.
11. A small unit of soldiers is called a ________________.
12. A ________________ is a small North American animal with a bushy striped tail.
13. Another name for a very violent windy storm is a ________________.
14. After a couple get married, they often go on a ________________.
15. A sailing ship with at least two masts is called a ________________.
16. If you are ________________ to a specific illness, you will not catch it.

F	M	N	L	A	G	O	O	N	T	P	R	K	P
O	M	P	R	U	N	E	K	S	Q	K	C	L	L
R	N	O	K	H	T	N	N	R	V	S	A	T	A
T	L	R	O	J	L	O	Q	C	N	M	R	H	T
U	G	M	W	N	A	Q	O	O	N	S	T	O	O
N	V	V	O	P	X	N	O	B	B	C	O	N	O
E	T	D	T	N	O	P	N	M	D	H	O	E	N
B	N	A	R	O	S	O	Y	E	L	O	N	Y	M
T	H	C	C	S	O	O	N	T	R	O	S	M	T
F	U	C	E	H	V	U	O	R	Z	N	L	O	S
Q	A	N	P	N	M	H	E	N	L	E	Q	O	E
R	U	Y	E	M	W	T	W	F	W	R	W	N	A
D	T	R	I	T	G	S	W	A	O	P	K	M	N
Y	N	R	B	A	L	L	O	O	N	S	M	M	R

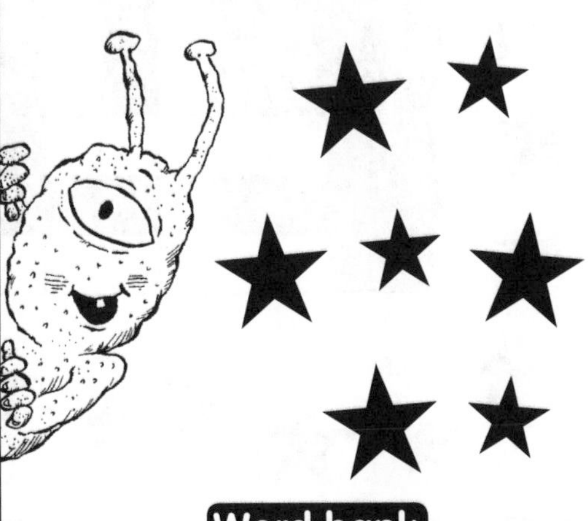

Word bank

BALLOONS	TYPHOON
SPOONS	HONEYMOON
MOON	SCHOONER
CARTOONS	DUNES
LAGOON	FORTUNE
MONSOON	IMMUNE
PLATOON	PRUNE
RACCOON	TUNE

Words with OOK

The letter string **OOK** is found in many words such as, *book*, *cook* and *hook*. The phoneme **OO** can sound short or long, for example, *food* (long) and *wood* (short). However, within the letter string **OOK**, it is different as it is always pronounced with a short **OO**. A rare exception to this rule is the word *spook*, which has a long **OO** phoneme. **OOK** can sometimes be pronounced in different ways according to regional dialects.

The words in this crossword all have the letter string **OOK** and a short **OO** phoneme.

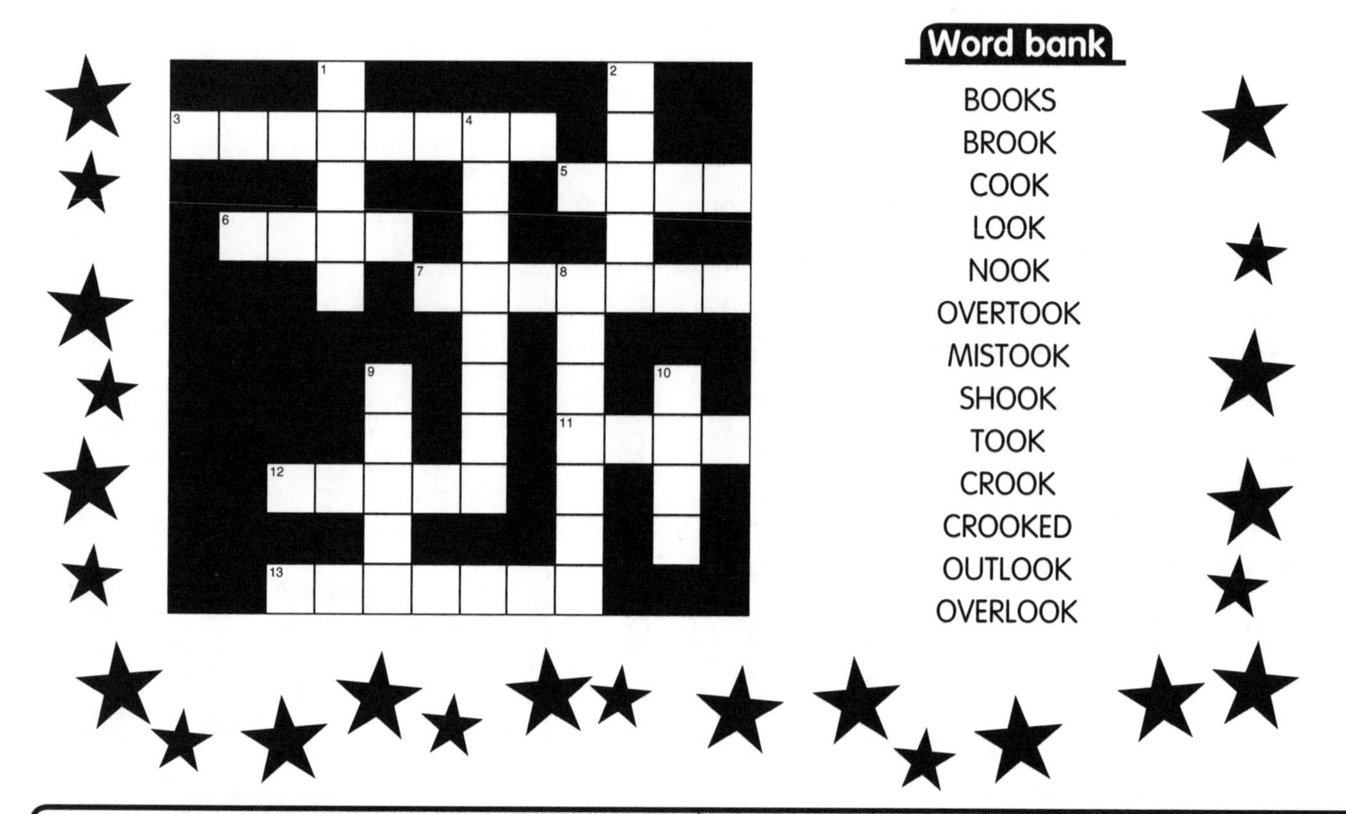

Word bank

BOOKS
BROOK
COOK
LOOK
NOOK
OVERTOOK
MISTOOK
SHOOK
TOOK
CROOK
CROOKED
OUTLOOK
OVERLOOK

ACROSS

3. The past tense of overtake (8)
5. Another name for a chef (4)
6. The past tense of take (4)
7. Not straight or level (7)
11. To use your eyes (4)
12. The past tense of shake (5)
13. The past tense of mistake (7)

DOWN

1. A small stream (5)
2. A criminal or a shepherd's curved staff (5)
4. To disregard or not notice (8)
8. A point of view (7)
9. Novels, encyclopaedias and dictionaries are all types of these (5)
10. A little sheltered corner or part of the expression '_ _ _ _ and cranny' (4)

Words with OW

The long **O** phoneme is heard in words with **OW**, **O**, **OE** and **OUGH**, for example, *window*, *radio*, *toe* and *though*. It can also be spelt **OH**, for example, *pharaoh*, but this spelling is far less common. The letter string **OW** can have different phonemes, for example, listen to the difference between *cow* – *snow* and *now* – *know*.

The words in this puzzle all contain **OW, OE, O** or **OUGH**. Complete the sentences using the word bank to help you, then find the words in the wordsearch.

1. The farmer put a ________________ in his field to frighten away the birds.
2. I weeded the garden and put all the waste in a ________________.
3. I stood ________________ the Eiffel Tower and looked up at it in amazement.
4. ________________ I had won the race I wasn't pleased with my time.
5. The explorers didn't want to catch malaria so they slept inside ________________ nets.
6. The sunflowers were a beautiful golden ________________ colour.
7. My gran moved into a ________________ so that she wouldn't have to use the stairs.
8. ________________ is the day after today.
9. At Christmas, people kiss under the ________________.
10. Country lanes can be really ________________ and not wide enough for two cars.
11. When my brother ran out of money, he tried to ________________ some from my mum.
12. I heard an ________________ as I called my name out in the cave.
13. When the avalanche began, the mountaineers tried to ________________ for help.
14. It was still raining but the sun came out and we saw a ________________.
15. When I kicked the ball, it went straight through the open ________________!

S	R	B	W	I	E	O	T	H	R	W	O	B
H	A	E	P	B	Z	L	G	N	O	H	W	N
O	I	L	Y	C	L	U	B	L	C	O	I	A
D	N	D	Y	E	O	G	A	E	R	H	N	R
V	B	O	B	H	L	G	E	C	L	L	D	R
G	O	W	T	N	N	L	E	J	B	E	O	O
N	W	L	M	U	W	R	O	Q	L	L	W	W
K	A	M	B	O	A	L	G	W	E	P	Z	F
G	M	D	R	C	N	K	M	X	T	W	R	V
Z	Z	R	S	M	I	S	T	L	E	T	O	E
M	O	S	Q	U	I	T	O	R	A	D	I	O
B	L	C	P	M	T	O	M	O	R	R	O	W
L	L	W	H	E	E	L	B	A	R	R	O	W

Word bank

BORROW	SCARECROW
RADIO	NARROW
MOSQUITO	ECHO
WHEELBARROW	WINDOW
BELOW	TOMORROW
BUNGALOW	RAINBOW
ALTHOUGH	YELLOW
MISTLETOE	

Words with OR and ORE

The phoneme **OR** can be spelt in a variety of ways. The different letter strings for this phoneme are found in hundreds of words. Some letter strings are more commonly used than others. Examples are as follows and are listed according to how commonly they are used:

- **OR**, for example, *for, mortar* and *ancestor.*
- **ORE**, for example, *bore, before* and *ignore.*
- **OUR**, for example, *your, source* and *detour.*
- **OAR**, for example, *boar, oar* and *roar.*
- **OOR**, for example, *floor, door,* and *poor.*
- **AUR**, for example, *dinosaur, minotaur* and *centaur.*
- **AR**, for example, *war* and *warning.*

Add the correct letter string (**OR**, **ORE**, **OUR**) to complete these words.

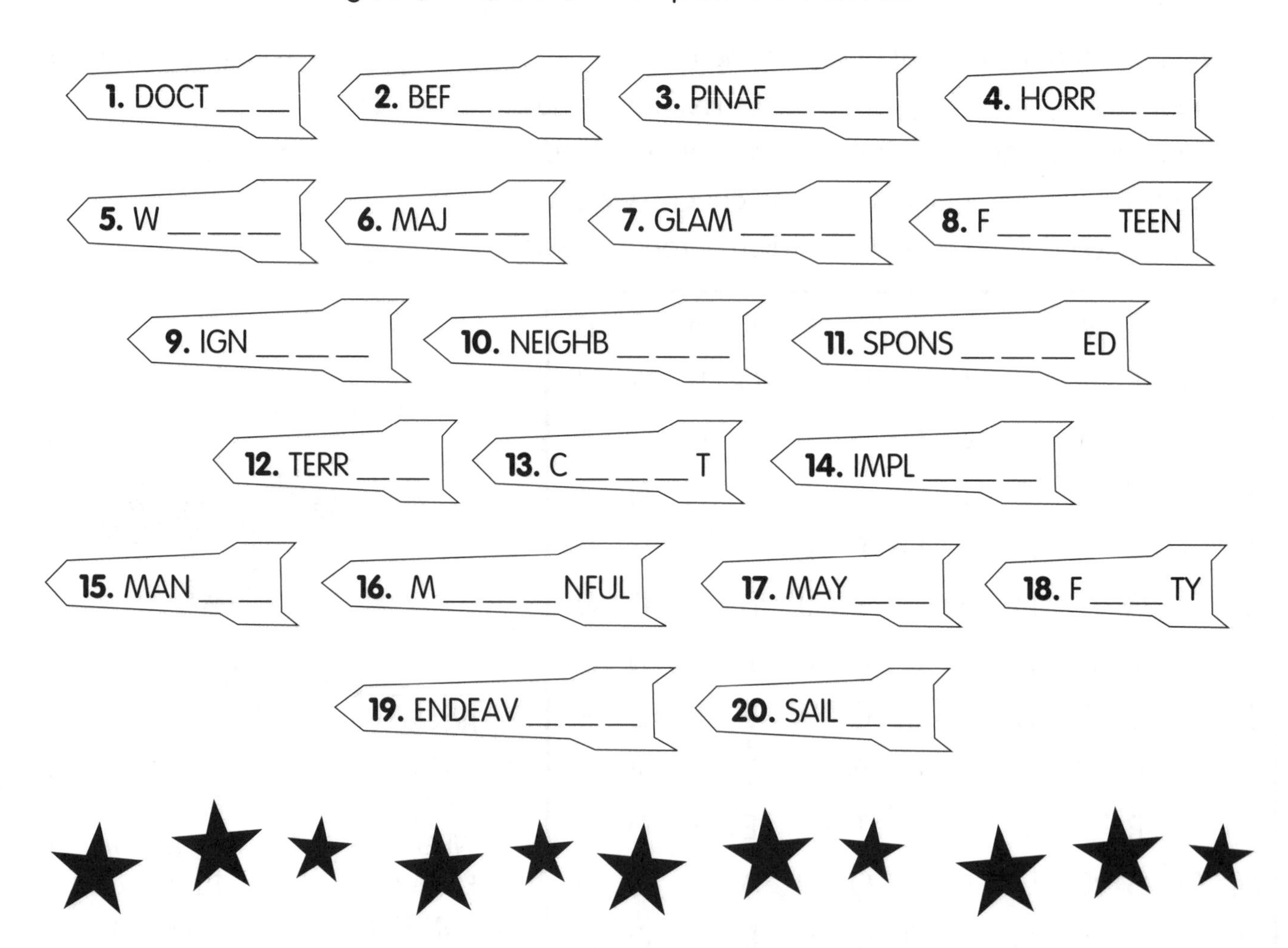

Words with OR and ORE

Sort these words according to how they sound and write them in the table below.

sailor normal your armour glamour fork adore scissors before
court conductor anchor comfort wore implore port endeavour
pork forty sponsored ignore doctor bore for major
bachelor cork course neighbour honour worm horn sculptor
protestor afford storm pour sore work world terror worth torn
tremor sport effort horror minor yore manor fourth mournful

ORE sounding like OR		ORE sounding more like ER	

★ ★

Sort these words according to their number of syllables and write them in the table below.

sport honour armour important dishonourable disorganised
conductor course commemoration organisation carnivorous sore
adore organise endeavour predecessor evaporation territorial
exploration professor ordinary sailor afford wore pour

1 syllable	2 syllables	3 syllables	4 syllables	5 syllables

★ ★

Words that end in ER

ER is a common final letter string. It can be part of a root word, for example, *brother, mother* and *slumber*, or it can be a suffix, added to words to make nouns, for example, *bake – baker, write – writer, run – runner* and *law – lawyer*. The rules for adding the suffix **ER** are the same as for adding **ED**. For words with:

- a final **E**, just add **R**, for example, *late – later*;
- two final consonants or a long vowel phoneme, just add **ER**, for example, *old – older*;
- a final **Y** after a consonant, change the **Y** to **I** and add **ER**, for example, *easy – easier*;
- a final **Y** after a vowel, keep the **Y** and add **ER**, for example, *play – player*;
- one final consonant with a short vowel phoneme, double the consonant and add **ER**, for example, *sad – sadder*.

You will find all types of words with **ER** in this crossword.

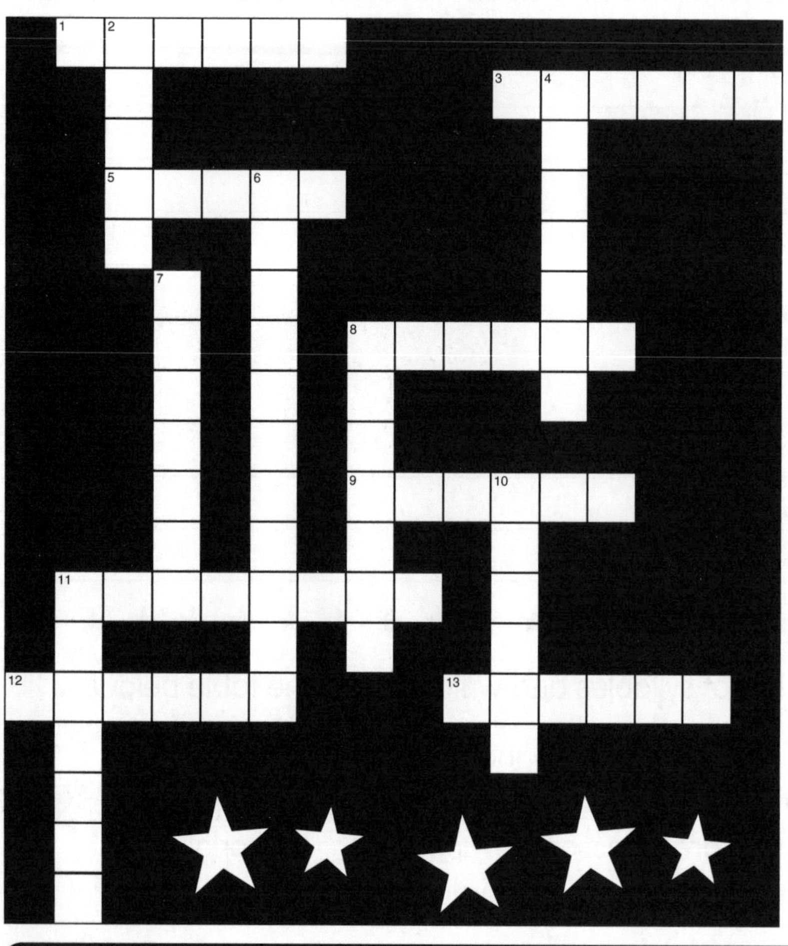

Word bank

STRANGER
LEARNER
ENTERTAINER
CORNER
BETTER
HOLSTER
SLOWER
TEACHER
TIDIER
ENTER
NEARER
CLEVER
STUTTER
LOSER
BUTTER

ACROSS

1. An antonym for faster (6) **3.** Intelligent or skillful (6) **5.** To go in (5) **8.** Neater (6) **9.** The place in a room where two walls meet (6) **11.** An unknown person (8) **12.** A spread made from churned milk (6) **13.** An antonym for worse (6)

DOWN

2. An antonym for winner (5) **4.** Someone who is being taught how to do something (7) **6.** Someone who performs for others amusement (11) **7.** A leather case for a gun, hung from a belt (7) **8.** Someone who works in a school (7) **10.** Closer not further away (6) **11.** To stammer and have trouble speaking (7)

Words that end in AR or OR

The words in the word bank all end in **AR** or **OR**. **OR** words are often nouns that describe a person who does something, for example, *collector, solicitor, dictator* and *operator*. **AR** words are less common. When **AR** is added to make adjectives, this suffix means 'having the nature of', for example, *muscular* and *solar*. Other **AR** words are nouns, for example, *collar, guitar* and *vinegar*.

The words in this crossword end in **AR** or **OR**.

Word bank

CALENDAR	TERROR
EDITOR	TRAITOR
AUTHOR	SAILOR
MUSCULAR	TARTAR
TRACTOR	MINOR
REGULAR	SOLAR
SPECTATOR	VISOR
ANCHOR	LUNAR
ACCELERATOR	RAZOR
SPONSOR	DONOR
MOTOR	ERROR

ACROSS

3. Consistent, usual or normal (7) **7.** To give money to a person or charity or to finance an event (7) **8.** The part of a car that makes it go, found under the bonnet (5) **10.** Press this foot pedal to make a car go faster (11) **11.** A person who writes books (6) **13.** A person who is in charge of a newspaper (6) **18.** A person who goes to sea in a boat (6) **19.** A heavy implement dropped in the water to stop a boat from drifting (6) **20.** A _ _ _ _ _ panel uses the sun's rays to make power (5) **21.** A crust on teeth (6)

DOWN

1. A mistake or misjudgement (5) **2.** Someone who watches something, for example, a sports match (9) **3.** A sharp instrument for shaving (5) **4.** Strong and lean (8) **5.** Someone who gives money or blood to others (5) **6.** A table of days and months in a year (8) **9.** Someone who betrays another or is guilty of treason (7) **12.** A vehicle used on a farm (7) **14.** The peak of a cap (5) **15.** Great fear or horror (6) **16.** Unimportant, small, not major (5) **17.** To do with the moon, for example, a _ _ _ _ _ eclipse (5)

Silent E

Each of these words contains a silent **E** which changes short vowels into long vowels, for example, *tap – tape, them – theme, pin – pine, hop – hope* and *us – use*. In most of the words, the silent **E** letter string is the final syllable but in some, the silent **E** letter string can be found in the first or second syllable, especially in compound words, for example, *wavelength, tasteless* and *gateway*. There are more letters strings with **A**, **I** and **O** combined with silent **E** than there are with **E** and **U**.

Find the eight silent **E** words in each letter puzzle and write a short definition for each word on a separate piece of paper.

Words with **A** + silent **E**:

c e n t i g r a d e a w a k e m u n d a n e i l l u s t
r a t e b e h a v e a m a z e m e n t e x h a l e l a c e

Words with **E** + silent **E**:

c e n t i p e d e t h e m e s c e n e t h e s e c
o m p l e t e t r a p e z e e v e n i n g p r e c e d e

Words with **I** + silent **E**:

a d v i c e d i v i d e l a n d s l i d e c r o c o d
i l e g r i m e c o m b i n e e x e r c i s e c a p s i z e

Words with **O** + silent **E**:

e x p l o d e g n o m e t a d p o l e p o s t p o n e t e
l e s c o p e c l o s e d c o v e f r o z e n

Words with **U** + silent **E**:

i n t r o d u c e h u g e d u k e m o l e c u l e
p e r f u m e t u n e f u l p i c t u r e i n j u r e

Silent E

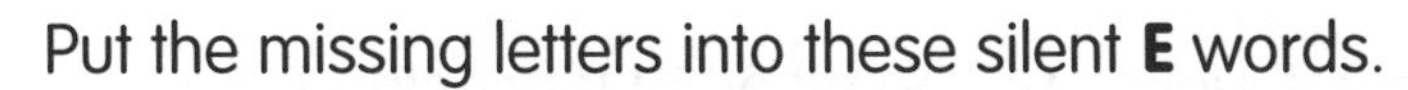

Put the missing letters into these silent **E** words.

	Definition	Incomplete word	Complete word
1	To focus and pay attention.	_ _ _ _ _ _ _ _ _ ATE	______________
2	Ten years.	_ _ _ ADE	______________
3	An antonym for demote.	_ _ _ _ OTE	______________
4	To eat or drink.	_ _ _ _ UME	______________
5	To get away or break free.	_ _ _ APE	______________
6	To come or go before.	_ _ _ CEDE	______________
7	The cure for poison or disease.	_ _ _ _ DOTE	______________
8	Keeping clean.	_ _ _ _ ENE	______________
9	To take part in competitions or contests.	_ _ _ _ ETE	______________
10	Two times.	_ _ ICE	______________
11	A young person.	_ _ _ _ _ ILE	______________
12	To be against someone or something.	_ _ _ OSE	______________
13	To sweat.	_ _ _ _ _ IRE	______________
14	A heavy fall of rain.	_ _ _ UGE	______________
15	An antonym for convex.	_ _ _ _ AVE	______________

Write in the correct silent **E** letter string to complete these words, for example, tadp *o l e*.

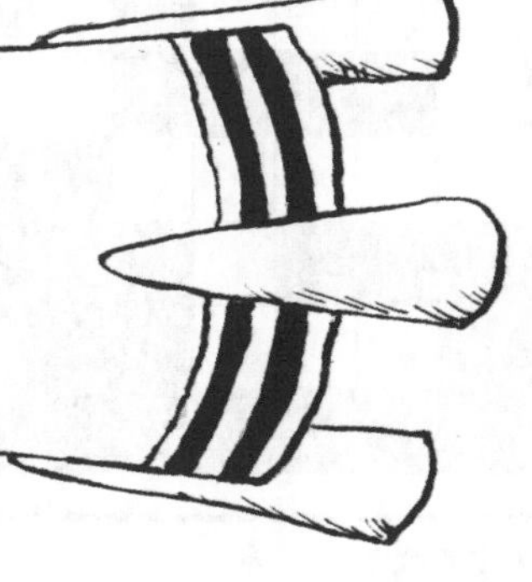

1. marmal _ _ _
2. earthqu _ _ _
3. vol _ _ _
4. stethosc _ _ _
5. sapph _ _ _
6. civil _ _ _
7. fem _ _ _
8. ridic _ _ _
9. cycl _ _ _
10. surn _ _ _
11. hurric _ _ _
12. advent _ _ _
13. Japan _ _ _
14. parad _ _ _
15. tessell _ _ _
16. grindst _ _ _
17. overt _ _ _
18. stalagm _ _ _
19. disp _ _ _
20. prosec _ _ _

Long A phoneme

The long **A** phoneme can be spelt in many different ways and the different letter strings for this phoneme can be found in hundreds of words. As well as occurring in root words, the long **A** phoneme can be found in suffixes, as outlined below. Some letter strings for long **A** are more common than others and some do not even include the letter **A**. Examples of the long **A** phoneme are:

- **A_E** is the most common, for example, *ate*;
- **AI**, for example, *tail*;
- **AY**, for example, *hay*;
- **A + LE** ending, for example, *table*;
- **A_**, for example, *apron*;
- **EI**, for example, *vein*;
- **EY**, for example, *they*.

Solve the clues to complete this crossword. Use the word bank to help you.

Word bank

FACE	THEY
PAID	WEIGHT
TASTE	DAY
STAIN	RAIN
STALE	TAME
PASTE	TAKE
LATE	MAY
HATE	MALE
WAIT	HAY
TAILS	SNAIL
AIM	MADE

ACROSS

1. An antonym for heads (5) **6.** To choose a direction to fire a weapon (3) **7.** Water falling from clouds (4) **11.** The past tense of make (4) **12.** The plural of he and she (4) **13.** Another word for flavour (5) **14.** Not on time (4) **15.** The past tense of pay (4) **16.** This can be measured in Kg or pounds and stones (6) **19.** The opposite of night (3) **20.** To delay in order for something to happen (4)

DOWN

1. The opposite of wild, like a pet (4) **2.** Not fresh (5) **3.** The front of the head (4) **4.** An antonym for female (4) **5.** A slow moving creature with a shell (5) **8.** An antonym for love (4) **9.** A mark or blemish you can't wash out (5) **10.** An antonym for give (4) **15.** A soft mixture (a type of glue perhaps) (5) **17.** Dried grass for animals to eat (3) **18.** It comes after April (3)

Long A phoneme

These words are just some of the thousands in the English language with long **A** phonemes. Sort them into groups below according to their letter strings. Some have been written in to help you.

displayed Australia mislay painting complaint agent misplace
famous mermaid payment Wednesday insulate illustrate veil
table eight ladle landscape bookcase masquerade alternating
disobeying Norway decade spacious surveyor subway sleigh
everyday again vacation alien location waste stable lazy
flame faint they survey playing danger pollinate reign fatal
nightingale Adrian phrase nature portray explain away delay
straying label communication raven opaque Asia April cradle

A_E (or A_E root word)	AI	A + single consonant	AY
vertebrate	waist	apron	display
	EY		
	obeyed		
	LE ending	**EI**	
	gable	reins	

Long E phoneme

The long **E** phoneme can be spelt in many different ways. Some letter strings are more common than others and some do not even include the letter **E**. Examples of the long **E** phoneme are:

- **EE**, for example, *tree*;
- **EA**, for example, *real*;
- **E_E**, for example, *scene*;
- **E**, for example, *be*;
- **E_** (but not at the end of words), for example, *secret*. (There are exceptions where the **E** phoneme is short, for example, *every*.);
- **IE** (according to the '**I** before **E**' rule), for example, *field*;
- **EI** (usually after by **C**), for example, *receive*. Some exceptions are *protein*, *weird*, *seize* and *weir*.

Read the clues and choose the correct letter string to complete the words.

	Clue	Choose EE, EA, IE, E_E or E
1	To contradict or say no.	DISAGR_ _
2	An antonym for war.	P_ _ CE
3	A bit, part or fragment.	P _ _ CE
4	A small segmented creature with many legs.	CENTIP _ D _
5	Nought.	Z _ RO
6	A plant that grows in the ocean.	S _ _ W _ _ D
7	A large bird of prey.	_ _ GLE
8	The ____________ champion is the overall winner.	SUPR _ M _
9	To become greater in size or number.	INCR _ _ SE
10	Short and concise.	BR _ _ F
11	An explorer or early settler.	PION _ _ R
12	An imaginary line of latitude around the middle of the Earth.	_ QUATOR
13	To hide.	CONC _ _ L
14	Vain or big-headed.	CONC _ _ TED
15	An antonym for shallower.	D _ _ PER

Long E phoneme

ACROSS

4. This can be single, double, whipped or clotted (5) **6.** The roof of a room (7) **9.** A person who takes part in sports that involve running, jumping and throwing (7) **11.** A fruit with a stone and a furry skin (5) **12.** The same in amount, size or value (5) **13.** My favourite _ _ _ _ _ _ is summer (6) **15.** To vanish from sight (9) **16.** When you buy something you may be given one of these as proof of payment (7)

DOWN

1. Seven days (4) **2.** Another name for a windscreen (10) **3.** Not finished (10) **5.** Films (6) **7.** Without worries or burdens (8) **8.** You can blow your nose in one of these (12) **10.** Shiny bright discs sewn on clothes to decorate them (7) **14.** Calm and tranquil (6)

Long I phoneme

The long **I** phoneme can be spelt in many different ways. Some letter strings are more common than others and some have the letter **Y** instead of **I**. Examples of the long **I** phoneme are:

- **I_E**, for example, *time*;
- **IGHT**, for example, *fight*;
- **IGH**, for example, *high*;
- **IE**, for example, *pie*;
- **IES** and **IED**, for example, *try – tries, tried*;.
- **ILD**, for example, *child*;
- **IND**, for example, *kind*;
- **I_LE**, for example, *idle*;
- **I_AL**, for example, *tidal*;
- **Y**, for example, *type*.

Underline the long **I** phonemes in these sentences.

1. The bride walked down the aisle of the church, wearing a light blue bridal gown.
2. The wild animals were sleeping in the shade as we tried to spy on them for a while.
3. At night they went hunting and travelled for miles across different tribal lands.
4. My little sister has a tricycle with wide tyres and I have a bicycle with nice shiny paintwork.
5. The man replied to the policeman's questions and denied all knowledge of the crimes.
6. We ignited a big bonfire and the flames shot high into the night sky, while the sparks began to fly like shooting stars.
7. I didn't have time to hide the prizes right away and the children started to fight over them.
8. He would have been released, but he was unwise and tried to bribe the police. They soon found out that he had lied.
9. The cowboy put the bridle on his ride, loaded his rifle, shined his grimy boots and galloped away, just in time.
10. The mountain was so high, we climbed for nine hours. My throat was dry and my thighs were tired.

Long I phoneme

Sort these words into groups in the table below. You can write them in more than one column if they belong in more than one group. Some words can be nouns and also present tense verbs, for example, *He is a **spy*** (noun) *and he likes to **spy*** (verb) *on people. There were lots of **spies*** (noun). *He **spies** on people* (verb).

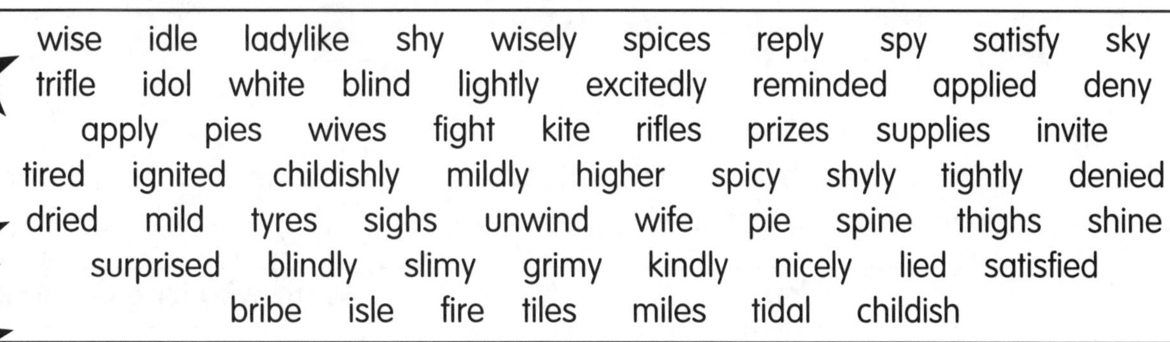

wise idle ladylike shy wisely spices reply spy satisfy sky
trifle idol white blind lightly excitedly reminded applied deny
apply pies wives fight kite rifles prizes supplies invite
tired ignited childishly mildly higher spicy shyly tightly denied
dried mild tyres sighs unwind wife pie spine thighs shine
surprised blindly slimy grimy kindly nicely lied satisfied
bribe isle fire tiles miles tidal childish

Noun	Noun – plural	Verb – present tense	Verb – past tense	Adverb	Adjective

Long O phoneme

The long **O** phoneme can be spelt in many different ways. Some letter strings are more common than others. Examples of the long **O** phoneme are:

- **O_E**, for example, *nose*;
- **OA**, for example, *boat*;
- **O_**, for example, *ocean*;
- **OW**, for example, *snow* (this letter string is not always pronounced with a long **O**, for example, *now*);
- **O**, for example, *go*;
- **OE**, for example, *toe*;
- **OLD**, for example, *old*.

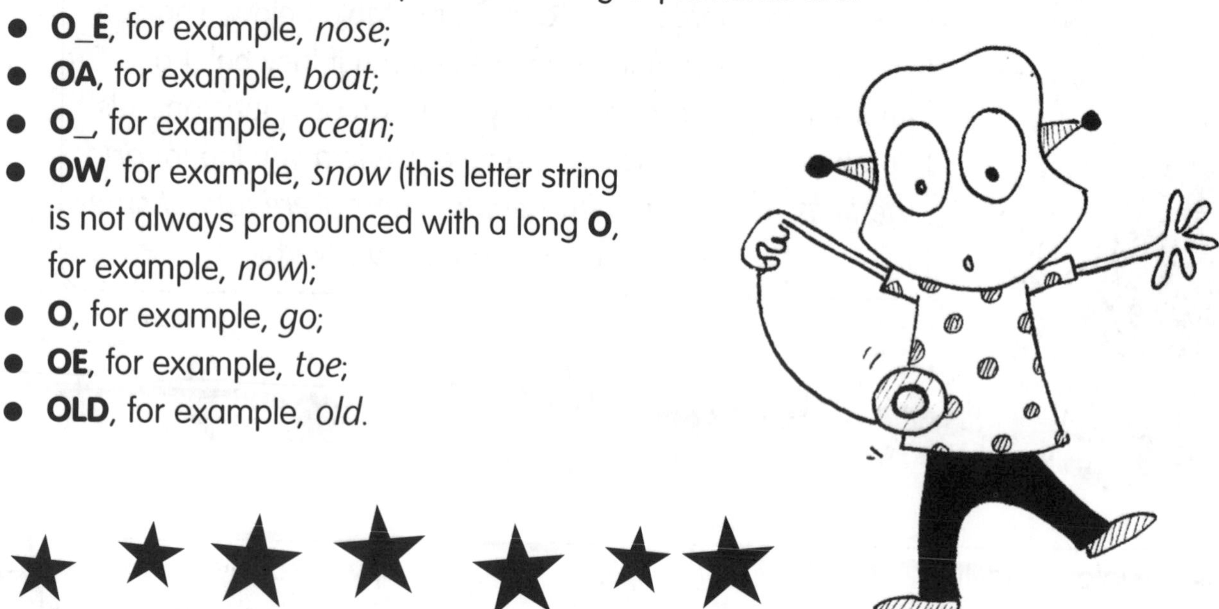

Look at the sets of words in the table. In the right-hand column, write the word from the set which has a long **O** phoneme. The first one has been done for you.

	Words	Word with long O phoneme
1	come, home, some	home
2	over, cover, hover	
3	town, own, brown	
4	road, abroad	
5	toe, shoe, canoe	
6	polish, Polish	
7	tomb, comb	
8	go, to	
9	should, shoulder	
10	bowl, fowl	
11	oven, woven	
12	cone, one	
13	approval, oval	
14	ove, drove	

Long O phoneme

Sort these words according to which syllables have a long **O** phoneme. If a word has two long **O** phonemes, then write it in both columns.

dynamo windows mistletoe photograph soapstone
hopeless postpone Apollo interrogate October solo
chocolate wardrobe overdose opponent broken
oboe tobacco episode inferno indigo

1st syllable	2nd syllable	3rd syllable

Read the clues and add the missing letters to these words with long **O** phonemes.

	Clue	Add the missing letters
1	A jacket or outer garment.	__ OA __
2	Sweet smelling flowers with sharp thorns.	__ O __ E __
3	A crease in a piece of paper or material.	__ OLD
4	Short for linoleum, a type of floor covering.	__ __ __ O
5	The bendy joints in your arms.	__ __ __ OW __
6	A very slow moving person.	__ __ OW __ OA __ __
7	A sieve is full of them.	__ O __ E __
8	What frogs do when they make a noise.	__ __ OA __
9	An aim, objective or ambition.	__ OA __
10	Not with anyone.	__ __ O __ E
11	Using oars in a boat.	__ OW __ __ __
12	An antonym for sink	__ __ OA __
13	Another word for the spine.	__ __ __ __ __ O __ E
14	To brag or 'blow your own trumpet'.	__ OA __ __
15	The colour of the rainbow between blue and violet.	__ __ __ __ __ O

Long U phoneme

The long **U** phoneme is not as common as long **A**, **E**, **I** or **O** and can be spelt in several ways. It can also be found in the suffix **DU**, for example, *duplicate*, as well as in root words. Examples of the long **U** phoneme are:

- **U_E**, for example, *cure;*
- **U**, for example, *unity;*
- **UA**, for example, *valuable;*
- **UE**, for example, *due;*
- **U_LAR**, for example, *regular;*
- **U_LATE**, for example, *calculate;*
- **U_LENT**, for example, *turbulent;*
- **U_LUS**, for example, *stimulus;*
- **EW**, for example, *few.*

These words have long **U** phonemes but unusual letter strings: *puma, tuna, duty, fury, beauty* and *beautiful.*

The first (underlined) words in these sets have long **U** phonemes and the others do not: *due* – *true, duke* – *Luke, union* – *bunion, cure* – *figure* and *astute* – *salute.*

Find the long **U** words in the grid. The letters are in the correct order from left to right, but the words have been muddled together. Write the words on the line. The first one has been done for you.

~~R~~	O	~~F~~	~~U~~	G	~~E~~	~~D~~		
S	O	R	U	U	N	A		
D	~~E~~	L	U	~~S~~	E	E		
P	E	R	T	B	M	E		
F	E	L	F	U	L	E	T	E

Example: Having said no to doing something – REFUSED

1. Sweet smelling scent – ______________
2. A flood or downpour – ______________
3. Able to be dissolved – ______________
4. Lucky – ______________

Join the long **U** homophone to the correct clues.

1. cue — A line of people.
queue — A long stick for playing snooker.

2. due — Expected.
dew — Tiny drops of water that form outside overnight.

3. new — Not old.
knew — Was aware of or had information about.

Long U phoneme

The words in this crossword all have long **U** phonemes.

Word bank

EXCUSE	CUBE
MUSEUM	CURE
OBTUSE	DUKE
CONFUSE	FUEL
MUSICAL	FUSE
NEPTUNE	HUGE
POPULAR	MULE
BARBECUE	MUTE
CUCUMBER	TUBE
PICTURES	DUNES
USE	TULIP

ACROSS
4. A long narrow cylinder (4) **6.** These can be framed and hung on the wall (8) **9.** Liked by a lot of people (7) **13.** A cross between a horse and donkey (4) **14.** Long green salad vegetable (8) **15.** A film or play with songs (7) **17.** To apply or operate (3) **18.** Sand hills near the beach (5) **20.** He's married to a duchess (4) **21.** A reason for behaviour (often bad) (6) **22.** A planet named after the Roman sea god (7)

DOWN **1.** An angle over 90 degrees (6) **2.** A solid shape with six square sides (4) **3.** A brightly coloured flower grown in Holland (5) **5.** Very large, enormous (4) **7.** To heal or make well again (4) **8.** A gallery of historical objects (6) **10.** A meal where meat is cooked outside (8) **11.** Petrol or gas (4) **12.** Part of an electrical system (4) **16.** To muddle or make unclear (7) **19.** Silent, unable to speak (4)

Plural nouns

Rules for plural nouns are:

- Add **S**, for example, *bag – bags*, *cake – cakes* and *taxi – taxis*.
- Add **ES** to words that end in **S**, **X**, **CH** and **SH**, for example, *pass – passes*, *box – boxes*, *church – churches* and *bush – bushes*.
- Change **F** to **V** and add **S** or **ES**, for example, *wife – wives* and *scarf – scarves*.
- Keep **F** if it follows two vowels especially **IE**, **EE** and **OO** or another **F**, for example, *beliefs*, *reefs*, *roofs* and *plaintiffs*.
- Change **Y** after a consonant to **IES** and add **S**, for example, *fly – flies*.
- Keep **Y** after a vowel and add **S**, for example, *day – days*.
- Add **ES** to words that end **O**, for example, *potato – potatoes*. (However, be aware though that there are many exceptions to this rule, for example, *avocados, stereos, pianos, radios, videos* and *studios*.)

- Irregular plurals (that do not follow the rules above): *Man – men, woman – women, child – children, foot – feet, tooth – teeth, person – people, goose – geese, ox – oxen, hoof – hoofs, mouse – mice, louse – lice, quiz – quizzes* and *loaf – loaves.*
- Irregular plurals (with foreign origins or Latin and Greek plural forms): *gateau – gateaux* (French), *appendix – appendices, cactus – cacti, crisis – crises, nucleus – nuclei, oasis – oases, thesis – theses* and *vortex – vortices.*
- Words with the same singular and plural forms: *sheep, fish, money* and *deer.*
- Words with no singular form (often because they come in pairs): *trousers, clothes, scissors, braces, pants, knickers, glasses, spectacles* and *cash.*

Complete this table by adding the missing plural or singular forms. (These words are all regular plurals that follow the rules, but there are some tricky ones too.)

	Singular form	Plural form		Singular form	Plural form
1	fox		11		scratches
2		toys	12	baby	
3	lunch		13		brushes
4	month		14	friend	
5	spy		15		displays
6		halves	16	scarf	
7	ash		17		pies
8		coughs	18		leaves
9		flies	19	toffee	
10		waves	20		caves

Plural nouns

Complete this table by adding the missing plural or singular forms. (Some of these words are irregular plurals that don't follow the rules.)

	Singular form	Plural form		Singular form	Plural form
1		men	11		gases
2		apples	12		shelves
3	chief		13	woman	
4		people	14		children
5	ear		15		potatoes
6		loaves	16		churches
7	foot		17	tooth	
8	piano		18	table	
9	dish		19	hoax	
10	fish		20		cacti

Make these nouns plurals. The letters are in the correct order from left to right, but the words have been muddled together. Write the words on the lines like the example.

V	H	L	L					
A	E	O	O	O	Y			
H	I	I	E	N	S	S	S	
L	A	R	L	E	G	Y	E	
C	P	C	M	E	E	I	S	S

Example: Valley – VALLEYS

1. apology – ________________
2. hero – ________________
3. louse – ________________
4. chimney – ________________

T	E	M	E	E					
G	O	U	E	Y	E				
B	L	L	R	S	O				
Y	U	S	S	S	S	L	S		
A	O	E	A	T	E	E	V	E	S

5. tomato – ________________
6. goose – ________________
7. bus – ________________
8. yourself – ________________
9. alley – ________________

Verbs with ED and ING

ED is added to a verb to change it to the past tense, for example, *She* ***looks*** *angry* (present tense) – *She* ***looked*** *angry* (past tense). **ING** can be added to verbs in the past, present or future tenses, for example, *I* ***am looking*** *for my friends* (present tense). *I* ***was looking*** *for my friends* (past tense). *I* ***will be looking*** *for my friends* (future tense).

Here are some rules for adding **ED** and **ING**. If the word has:

- a short vowel phoneme and one syllable, ending in a single consonant, double the last letter and add **ED** or **ING**, for example, *hop – hopped/hopping;*
- a long vowel phoneme and one syllable, ending in a single consonant, add **ED** or **ING**, for example, *dream – dreamed/dreaming;*
- a short vowel phoneme and more than one syllable, ending in a single consonant, add **ED** or **ING**, for example, *open – opened/opening;*
- a short vowel phoneme and one syllable, ending in two consonants, add **ED** or **ING**, for example, *jump – jumped/jumping;*
- the last letter **E**, take off the **E** and add **ED** or **ING**, for example, chase – chased/chasing;
- the last letter **Y** after a vowel, add **ED** or **ING**, for example, *play – played/playing;*
- the last letter **Y** after a consonant, change the **Y** to **I** and add **ED**, for example, *spy – spying;* the last letter **Y** after a consonant, just add **ING**, for example, *spy – spying.*

Complete this table by adding the missing words.

Verb	Verb with ED	Verb with ING
wait		
		grumbling
happen		
borrow		
		drying
	exploded	
stray		
		screaming
		scratching
enjoy		
		illustrating
	stirred	
tour		

Hard and soft C and G

The rules for hard and soft **C** and **G** are:

- The letter **C** is **soft** when followed by **E**, **I** or **Y**, for example, *ceiling*, *city* and *cyclist*. (There are no exceptions to this rule.)
- The letter **G** is also **soft** when followed by **E**, **I** or **Y**, for example, *genes*, *giant* and *gypsy*. However, there are several exceptions to this rule, especially when **G** is followed by **I** and a long **E** or **O** phoneme, for example, *gift*, *giggle*, *gear*, *goose* and *geese*.

C and **G** are both **hard** when:

- followed by **A**, **O** or **U**, for example, *medical*, *collar*, *curve*, *galleon*, *golden*, *gulls* and *guide*;
- followed by any consonant, for example, *clown*, *glow*, *crisp*, *grab*, *act* and *back*. (**C** followed by **H** can have different phonemes, for example, *church*, *chair*, *Christmas* and *choir*.);
- found on their own at the end of words, for example, *panic*, *music*, *attic*, *electric*, *picnic*, *bag*, *earwig*, *whirligig* and *hedgehog*.

The following words have letter strings that look the same but they actually have different phonemes. Read them out loud to hear the difference: *danger – hanger*, *finger – ginger*, *gibbon – giblets* and *girl – giro*.

Sort these words according to their **C** and **G** phonemes and write them in the table below.

occurring bigger special fragile large spaceman collar
magical curtains gasping Cyprus Cinderella magician politician
original bingo pelican ridiculous goalposts catapult goose
regularly receive certainly generally giraffe gymnastics local
imagination goggles dangerous galleon

Hard C	Hard G	Soft C	Soft G

Words with GH and GHT

GH and **GHT** can be in the middle or at the end of words. They can follow various vowel combinations to create many different phonemes as listed below. Sometimes you cannot even hear that they are there!

The letter strings **IGH**, **IGHT AUGH** and **EIGH** are the easiest to remember:

- **IGH** – long **I** phoneme, for example, *sigh*;
- **IGHT** – long **I** phoneme, for example, *sight*;
- **EIGH** – long **A** phoneme, for example, *weigh*;
- **AUGH** – short **A** phoneme, for example, *laugh*;
- **EIGHT** – long **A** phoneme, for example, *weight* (Exception: long **I** phoneme, for example, *height*;
- **AUGHT** – **AU** digraph, for example, *caught*.

Word bank

ROUGH
LIGHTEST
FLIGHT
COUGH
TIGHT
THOUGHT
THROUGH
NEIGHBOUR
MIDNIGHT
FRIGHT
HIGH
NOUGHT
SIGH
THIGH
KNIGHT
FIGHT
TAUGHT
HEIGHT
DRAUGHT

The words in this crossword all contain **GH** and **GHT**.

ACROSS

1. An antonym for loose (5) **4.** An antonym for heaviest (8) **7.** A chess piece or a medieval man in armour (6) **10.** To compete in battle (5) **14.** A current of cold air usually indoors (7) **15.** A long audible breath (4) **16.** Not smooth (5) **17.** Zero (6) **18.** The upper part of the leg (5) **19.** Sudden fear or alarm (6)

DOWN

2. The measurement of how tall something is (6) **3.** Movement through the air or the name for a set of stairs (6) **5.** Threw (homophone) (7) **6.** Twelve o'clock at night (8) **8.** A person who lives nearby (9) **9.** The past tense of think (7) **11.** The past tense of teach (6) **12.** Not low (4) **13.** To clear your throat and expel air from the lungs (5)

I before E except after C

The vowel digraphs **IE** and **EI** can sound the same. When they create a long **E** phoneme, the **I** comes before the **E**, for example, *belief*, *chief* and *field*. However, if these letters follow **C**, then they are reversed and written **EI**, for example, *conceit*, *receive* and *ceiling*. Exceptions to the '**I** before **E**' rule are *protein*, *seize*, *weir*, *perceive*, *weird* and *caffeine*. The word *view* is also an exception as it has a long **U** sound. The words *neither* and *either* are often pronounced in different ways according to regional dialects.

Sort these words with **IE** and **EI** according to their phoneme in the table below. Some follow the rules and some are exceptions.

kaleidoscope freight reign seize belief Fahrenheit weigh
neighbour brief hygienic poltergeist yield eiderdown height
seismograph weir retriever series protein shriek fiend pierce
grieve sleigh reindeer vein weight reins heir masterpiece
achieve deceit conceited receive weird niece siege perceive veil

Long E		**Long A**	**Long I**

Correct the wrongly spelt words in this list and write them on the lines. Tick the correctly spelt words.

1. HANDKERCHIEF ____________________

2. RECIEPT ____________________

3. SLEIGH ____________________

4. PEIRCE ____________________

5. ACHEIVEMENT ____________________

6. FRIEGHTER ____________________

7. FREIND ____________________

8. REPREIVE ____________________

Suffixes – changing nouns and adjectives into verbs

The **suffixes** in these puzzles change **nouns** and **adjectives** into **verbs**. The table below lists the four most commonly used suffixes that can do this. The endings **ICE** and **ISE** can sometimes be confusing. The general rule is that if a noun ends in **ICE**, for example, *advice*, then the verb will end in **ISE**, for example, advise.

Complete the table. The first one has been done for you.

Suffix	Noun	Verb
ISE	idol	idolise
	victim	
ATE	liquid	
	migrant	
IFY	class	
	clear	
EN	fright	
	wide	

Add the correct suffixes to the words in the word bank to change them into verbs. Then highlight these verbs in the wordsearch. One has been done for you.

A	R	L	P	M	K	C	R	I	T	I	C	I	S	E	W
C	D	E	B	A	D	N	D	I	V	E	R	S	I	F	Y
W	R	V	G	R	N	K	E	M	P	H	A	S	I	S	E
S	M	T	E	U	I	A	C	E	L	E	B	R	A	T	E
N	O	Z	C	R	L	G	L	D	E	M	O	N	I	S	E
X	Z	F	K	S	T	A	H	Y	P	W	T	F	E	R	E
I	P	E	T	P	W	I	T	T	S	C	L	T	D	S	D
D	N	N	F	E	D	P	S	E	E	E	A	Y	I	L	E
E	N	E	R	C	N	F	J	E	G	N	F	L	K	T	S
N	A	R	I	I	X	T	N	M	G	I	A	E	A	L	H
T	R	G	G	A	B	B	H	I	D	D	S	U	N	K	A
I	R	I	H	L	W	K	S	I	N	I	T	E	R	L	R
F	A	S	T	I	R	E	L	A	V	C	D	T	F	Q	P
Y	T	E	E	S	D	O	V	E	N	R	P	D	T	W	E
L	E	Z	N	E	S	G	D	U	A	R	N	X	K	N	N
F	A	L	S	I	F	Y	P	H	J	P	U	R	I	F	Y

Word bank

~~DEVISE~~
CRITIC
ANALYSIS
SPECIAL
VANDAL
EMPHASIS
ADVERT
NARRATOR
PUNCTUATION
CELEBRATION
REGULAR
FRIGHT
SOLID
FALSE
PURE
BRIGHT
SHARP
HARD
SOFT
IDENTITY
DIVERSITY
DESIGN
DEMON
ENERGY

Changing verbs into nouns

The **suffixes** on this page can change **verbs** into **nouns**. The table below lists the most commonly used suffixes that can do this. Complete the table. The first one has been done for you.

Suffix	Verb	Noun
ADE	block	blockade
AGE	block	
AL	refuse	
ANCE	assist	
ANCY	occupy	
ANT	occupy	
AR	beg	
DOM	free	
EE	employ	
ENT	correspond	
ER	teach	
ION	infect	
MENT	astonish	
OR	direct	

Using the suffix list to help you, add the correct suffixes to the verbs in the clues and complete the crossword. One has been done for you.

ACROSS
- ~~2. Refer (7)~~
- **7.** Drain (8)
- **8.** Bore (7)
- **10.** Move (8)
- **15.** Assist (9)
- **16.** Escape (8)
- **17.** Farm (6)
- **18.** Allow (9)
- **19.** Act (6)
- **20.** Enjoy (9

DOWN
- **1.** Decide (8)
- **3.** Free (7)
- **4.** Narrate (9)
- **5.** Leak (7)
- **6.** Account (10)
- **9.** Rehearse (9)
- **11.** Post (7)
- **12.** Amaze (9)
- **13.** Spectate (9)
- **14.** Vacate (7)

Adjectives and adverbs

Adjectives give more information about nouns, for example, *A **blue** car*. **Adverbs** are words that give more information about verbs, for example, *I run **quickly***. Adjectives can often be changed into adverbs by adding the suffix **LY**, for example, *quick – quickly*. Here are some rules:

- If the adjective has more than one syllable and ends in **Y**, then the **Y** is replaced with **I**, for example, *happy – happily*.
- If the adjective ends in **LE**, remove the **E** before adding **LY**, for example, *gentle – gently*.
- If the adjective ends in **IC**, then **ALLY** must be added, for example, *energetic – energetically*.
- Not all adjectives can be changed into adverbs and **LY** cannot be added to colours and numbers.

ACROSS

4. Bright (change to an adjective) (8) **6.** Angry (change to an adverb) (7) **10.** Energetic (change to an adverb) (13) **12.** Hopeless (change to an adverb) (10) **13.** This adjective is in between first and third (6) **14.** Wicked (change to an adverb) (8)

DOWN

1. Fun (change into adverb) (7) **2.** Heroic (change to an adverb) (10) **3.** An antonym for weakly (8) **5.** Public (change to an adverb) (8) **7.** Blue and yellow mixed together make this adjective (5) **8.** Friend (change to an adjective) (11) **9.** Realistically (change to an adjective) (9) **11.** A synonym for intelligently (8)

Suffixes

The suffixes **CEDE**, **CLUDE**, **LOGUE**, **LOGY**, **SCRIBE** and **SCOPE** all provide additional information about the words that are in front of them and have Greek or Latin roots. This table gives some examples:

Latin roots	Greek roots
CEDE: concede, precede	**LOGY**: trilogy, eulogy
CLUDE: conclude, include	**LOGUE**: catalogue, dialogue
SCRIBE: inscribe, describe	**SCOPE**: telescope, microscope

In the table below, the words (with suffixes) and their definitions have been muddled up. Write the correct word next to each definition. One has been done for you.

Word	Definition	Correct word
MONOLOGUE	To set out rules, or order the use of certain medicines.	PRESCRIBE
~~PRESCRIBE~~	A study or an account of a family tree.	
EPILOGUE	An instrument used in a submarine or to give views of things on different levels.	
INCLUDE	A long speech by one person.	
GENEALOGY	To say what someone or something is like.	
PERISCOPE	A closing speech.	
TRILOGY	A series of three related works or stories.	
EXCLUDE	To involve, add in or make part of.	
DESCRIBE	A tube that you look through to see brightly coloured changing patterns.	
KALEIDOSCOPE	To come or go before or in front of something.	
PRECEDE	To shut or keep someone or something out.	

Each of the six words in this letter puzzle has one of the suffixes listed at top of the page. Circle the words and write a short definition or a sentence for each one on a separate piece of paper. The first one has been done for you.

dial ogu einter cedec onclud este tho sco peeul ogytr ans cri be

Suffixes revision

The words in this table contain suffixes that you have met in previous spelling work. The letters in brackets tell you what type of word usually has this suffix: N = noun, V = verb, Adj = adjective, Adv = adverb. When a suffix is added, a word can often change type, for example, *refuse* (V) – *refusal* (N); *help* (V or N) – *helpful* (Adj); *loud* (Adj) – *loudly* (Adv) and *beauty* (N) – *beautify* (V).

ABLE (Adj) reliable	**AL** (N) refusal	**ARY** (Adj/N) legendary, dictionary	**ATE** (V) translate	**CEDE** (V) precede	**CIAN** (N) magician
CLUDE (V) include	**EN** (V) darken	**ESS** (N) princess	**ETTE** (N) cigarette	**FUL** (Adj) hopeful	**HOOD** (N) falsehood
IAN (N) librarian	**IBLE** (Adj) horrible	**IFY** (V) horrify	**ION** (N) revision	**ISE** (V) televise	**ISM** (N) mechanism
LIKE (Adv) childlike	**LOGY** (N) analogy	**LY** (Adv) quickly	**OGUE** (N) dialogue	**LESS** (Adj) careless	**MENT** (N) amusement
NESS (N) calmness	**OLOGY** (N) biology	**SCOPE** (N) telescope	**SCRIBE** (V) transcribe	**SHIP** (N) friendship	**WORTHY** (Adj) roadworthy

	Word	Word with correct suffix
1	comfort	comfortable
2	concentr	
3	terr	
4	con	
5	forbidd	
6	beaut	
7	histor	
8	revers	
9	cass	
10	volunt	
11	politi	
12	pre	
13	cheer	
14	brother	
15	lion	

These 15 words have the following suffixes: **ABLE, AL, ARY, ATE, CEDE, CIAN, CLUDE, EN, ESS, ETTE, FUL, HOOD, IAN, IBLE, IFY.** Write the words on the lines with the correct suffixes. The first one has been done for you.

Suffixes revision

	Word	Word with correct suffix
1	metabol	
2	tri	
3	real	
4	catal	
5	decis	
6	fault	
7	lady	
8	amuse	
9	faithful	
10	chron	
11	aggressive	
12	in	
13	kin	
14	praise	
15	tele	

These 15 words have the following suffixes: **ION, ISE, ISM, LIKE, LOGY, LY, OGUE, LESS, MENT, NESS, OLOGY, SCOPE, SCRIBE, SHIP, WORTHY**. Write the words on the lines with the correct suffixes.

You can read the suffixes for these words, but the first letters are jumbled up. Put them in the correct order and write the complete words in the spaces. The first one has been done for you.

1.	newer **able**	*renewable*	16.	shinotsa **ment**	
2.	prevus **ise**		17.	zag **ette**	
3.	green **ate**		18.	hutgoth **fully**	
4.	elu **ogy**		19.	dubon **ary**	
5.	xelf **ible**		20.	het **ology**	
6.	creex **ise**		21.	tricceel **ian**	
7.	xe **clude**		22.	hys **ness**	
8.	pile **ogue**		23.	trine **cede**	
9.	zorf **en**		24.	ed **scribe**	
10.	planetxa **ion**		25.	butdo **ful**	
11.	gamn **ify**		26.	dhar **ship**	
12.	mobtot **less**		27.	dilch **hood**	
13.	squeert **ian**		28.	mealb **worthy**	
14.	flie **like**		29.	triwa **ess**	
15.	vinesur **al**		30.	rocmi **scope**	

Prefixes

The prefixes **AERO**, **AQUA**, **CO/CON**, **CONTRA**, **CRED**, **DU** and **HYDRO/HYDRA** all provide additional information about the words that follow them and have Greek (G) or Latin (L) roots. This table gives some examples:

AERO: aeroplane, aerosol (G)	**CONTRA**: contrast, contradict (L)
AQUA: aquatic, aquarium (L)	**CRED**: credible, credit (L)
AUDI: audible, audience (L)	**DU**: duplicate, duet (L)
CO/CON: concave, coincidence (L)	**HYDRO/HYDRA**: hydrofoil, hydrogen, hydrate (G)

In the table below, write in the missing letters to complete the words.

Word	Definition
AERO _ _ _ _ _ _	Stunts carried out in the air.
CON _ _ _ _	To hide.
CRED _ _ _ _ _ _ _	Authorisation or someone's personal details or qualifications.
AERO _ _ _ _	Exercises that strengthen your heart and lungs.
_ _ _ _ _ ELECTRICITY	Power created by water.
_ _ _ _ OVISUAL	Information delivered with sound and pictures.
_ _ _ _ _ _ DICT	To say the opposite or that someone or some information is wrong.
_ _ _ CLUDE	To end.
_ _ _ _ TION	A test to see if a performer is suitable.
_ _ _ _ ITABLE	Trustworthy and honest.
AQUA _ _ _ _ _ _	A bluish-green colour.
CONTRA _ _	Obstinate or awkward, deliberately doing the opposite.
_ _ EL	A fight between two people.
_ _ _ _ RIUS	A sign of the zodiac – the water bearer.
DU _ _ _ _ _ _ _	Double dealing and dishonesty.
HYDRA _ _	A connection for attaching a hose to the water mains.

Prefixes

The prefixes **MICRO**, **OCT**, **PHOTO**, **PORT**, **PRIM**, **SUB**, **TRI** and **EX** all provide additional information about the words that follow them and have Greek (G) or Latin (L) roots. This table gives some examples:

EX: external, ex-army (L)	**SUB**: substitute, subway (L)
PORT: portcullis, portable (L)	**OCT**: octopus, octagon (G)
TRI: tricycle, trio (L/G)	**PRIM**: primary, prima donna (L)
PHOTO: photocopy (G)	**MICRO**: microclimate, microphone (G)

In the table below, write in the missing letters to complete the words.

Word	Definition
_ _ _ AVE	The range of eight notes.
PHOTO _ _ _ _ _ _	A very close finish that needs a camera to help the decision.
SUB _ _ _ _ _ _ _ _ _ _	Underground.
OCT _ _	A group of eight musicians.
MICRO _ _	An extremely small organism too small for the eye to see.
_ _ _ _ ER	Someone who looks after a hotel lobby or carries guests' bags.
_ _ _ MERGE	To put under the surface of liquid.
EX _ _ _ _ _ _	To hurry or speed up a process.
_ _ _ _ _ WAVE	A cooking device that uses energy in very short waves.
_ _ _ _ HOLE	A small window in the side of a ship.
TRI _ _ _ _ _ _ _	Able to speak three languages.
_ _ _ _ _ FOIL	A light vessel with a hull that raises out of the water at speed.
_ _ _ _ _ SYNTHESIS	The process by which plants turn CO_2 and water into energy.
_ _ _ _ _ SCOPE	An instrument for magnifying minute objects.
_ _ TERIOR	The outside.
_ _ _ ANGLE	A 2-D shape with three sides.

Prefixes revision

This wordsearch contains 16 words, one for each of the following prefixes: **AERO**, **AQUA**, **ANTI**, **AUTO**, **BI**, **CIRCUM**, **CO**, **CON**, **CONTRA**, **CRED**, **DE**, **DIS**, **DU**, **EX**, **HYDRO** and **HYDRA**. Read the clues carefully and think which prefix may apply, then circle it in the wordsearch. Use the word bank to help you.

1. A substance used to reduce sweating (14) **2.** A signature (9) **3.** To divide into two equal parts (6) **4.** To sail right around (14) **5.** To organise and bring into order (11) **6.** A face-to-face conflict or contest (13) **7.** Movement of traffic in the opposite direction (10) **8.** Believable (8) **9.** To let the air or gas out of something (7) **10.** Not pleased (12) **11.** To make an exact copy (9) **12.** To speak suddenly and cry out (7) **13.** An aversion or fear of water (11) **14.** Something operated by pressure transmitted by liquid in pipes (9) **15.** Another name for a spray can (7) **16.** A building with large fish tanks (8)

M	W	R	Z	R	K	V	Y	D	K	Q	F	A	M	R	B
C	R	A	N	T	I	P	E	R	S	P	I	R	A	N	T
Y	O	J	D	T	H	M	A	Q	B	B	B	E	C	F	T
M	C	N	G	J	G	D	L	Q	O	R	T	G	L	R	Z
L	R	V	T	Z	P	N	U	H	U	A	K	O	K	E	B
M	E	H	D	R	M	G	P	P	N	A	S	Z	T	J	B
C	D	K	N	G	A	O	B	I	L	O	R	A	Z	C	E
A	I	G	L	Q	R	F	D	I	R	I	L	I	W	Z	X
U	B	K	H	D	L	R	L	E	S	F	C	Q	U	P	C
T	L	T	Y	N	O	V	A	O	E	E	V	A	C	M	L
O	E	H	P	-	C	X	G	D	W	B	C	J	T	L	A
G	Q	N	O	V	W	T	L	Z	T	Y	N	T	L	E	I
R	T	C	O	N	F	R	O	N	T	A	T	I	O	N	M
A	N	D	I	S	S	A	T	I	S	F	I	E	D	R	N
P	D	C	I	R	C	U	M	N	A	V	I	G	A	T	E
H	Q	N	H	Y	D	R	A	U	L	I	C	N	R	Q	J

Word bank

AQUARIUM
HYDRAULIC
ANTIPERSPIRANT
AUTOGRAPH
BISECT
CIRCUMNAVIGATE
CONFRONTATION
CREDIBLE
DEFLATE
DISSATISFIED
DUPLICATE
EXCLAIM
AEROSOL
HYDROPHOBIA
CO-ORDINATE
CONTRAFLOW

Prefixes revision

This crossword contains 19 words, one for each of the following prefixes: **IL, IM, IR, MICRO MIS, NON, OCT, PHOTO, PORT, PRE, PRIM, PRO, RE, SUB, SUS, TELE, TRANS, TRI** and **UN**. Read the clues carefully and think which prefix may apply. Use the word bank to help you.

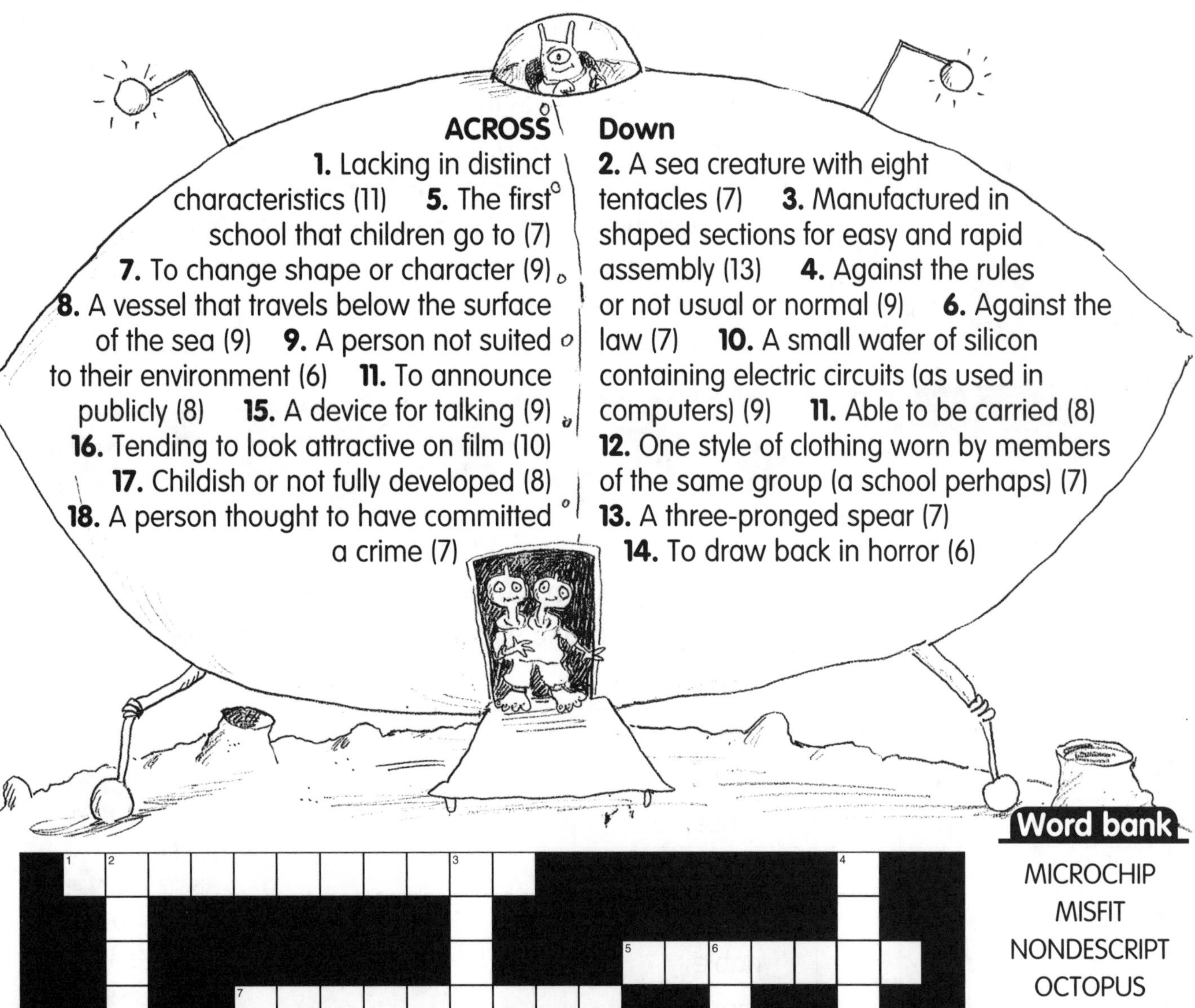

ACROSS

1. Lacking in distinct characteristics (11) **5.** The first school that children go to (7) **7.** To change shape or character (9) **8.** A vessel that travels below the surface of the sea (9) **9.** A person not suited to their environment (6) **11.** To announce publicly (8) **15.** A device for talking (9) **16.** Tending to look attractive on film (10) **17.** Childish or not fully developed (8) **18.** A person thought to have committed a crime (7)

Down

2. A sea creature with eight tentacles (7) **3.** Manufactured in shaped sections for easy and rapid assembly (13) **4.** Against the rules or not usual or normal (9) **6.** Against the law (7) **10.** A small wafer of silicon containing electric circuits (as used in computers) (9) **11.** Able to be carried (8) **12.** One style of clothing worn by members of the same group (a school perhaps) (7) **13.** A three-pronged spear (7) **14.** To draw back in horror (6)

Word bank

MICROCHIP
MISFIT
NONDESCRIPT
OCTOPUS
PHOTOGENIC
PORTABLE
PREFABRICATED
PRIMARY
PROCLAIM
RECOIL
SUBMARINE
SUSPECT
TELEPHONE
TRANSFORM
TRIDENT
UNIFORM
ILLEGAL
IMMATURE
IRREGULAR

Misspelt words with unstressed vowels

Some words are tricky to spell correctly because they contain unstressed vowels. These vowels are spoken quietly or quickly or they do not sound out clearly because the stress is on another syllable in the word, for example, *everywhere*, *different* and *around*. In these words it can be hard to hear whether the vowel is **A**, **E**, **I**, **O** or **U**, or whether there is a vowel there at all. Some words are also hard to spell correctly because of the inexact way we often pronounce them, for example, *doctor*, *library* and *probably*.

Choose an unstressed vowel or vowel diagraph (**a**, **e**, **io**, **o** or **au**) and place it in the column between the two parts of a word. Write the completed word in the final column. The first one has been done for you.

1	pref	e	rable	preferable
2	math		matics	
3	ref		rence	
4	desp		rate	
5	valu		ble	
6	fash		nable	
7	rest		rant	
8	temp		rament	
9	cemet		ry	
10	tempor		ry	
11	laborat		ry	
12	vet		rinary	
13	libr		ry	
14	magic		lly	

Misspelt words with unstressed vowels

Some of these words are spelt wrongly and some are spelt correctly. Tick those that are spelt correctly and write the correct spelling next to the words which are misspelt, then highlight them in the wordsearch.

1. TEMPERTURE ____________________
2. PROBLY ____________________
3. TOWL ____________________
4. EVRYWHERE ____________________
5. SOMEONE ____________________
6. ENERGETICLY ____________________
7. ENTHUSIASTICALLY ____________________
8. HAPPNING ____________________
9. EXTRORDINARY ____________________
10. FASHIONABLE ____________________
11. VOLUNTRY ____________________
12. MINIATURE ____________________
13. MISRABLE ____________________
14. VEGTABLE ____________________
15. JEWELLRY ____________________
16. TEMPORARY ____________________
17. BUSNESS ____________________

N	G	T	F	P	R	O	B	A	B	L	Y	P	C	G	X	P
E	P	E	T	W	R	H	T	B	P	L	B	M	G	J	W	K
N	Z	M	M	N	E	F	A	M	U	G	M	E	C	T	J	V
T	M	P	M	Z	V	L	V	P	R	S	L	R	E	L	E	V
H	I	O	I	R	E	L	T	C	P	B	I	N	Z	R	R	E
U	N	R	S	F	R	V	L	O	A	E	O	N	U	V	L	T
S	I	A	E	F	Y	M	O	T	W	E	N	T	E	B	J	W
I	A	R	R	R	W	N	E	L	M	E	A	I	A	S	V	Z
A	T	Y	A	B	H	G	Y	O	U	R	L	N	N	L	S	N
S	U	G	B	K	E	X	S	R	E	N	O	K	J	G	J	K
T	R	D	L	V	R	K	T	P	R	I	T	Z	Z	W	P	M
I	E	X	E	R	E	T	M	Z	H	K	R	A	K	G	N	K
C	D	C	G	W	T	E	W	S	Z	Y	P	N	R	H	T	T
A	M	F	F	P	T	K	A	K	N	K	R	N	T	Y	Q	P
L	D	D	J	F	Y	F	J	E	W	E	L	L	E	R	Y	Z
L	L	K	E	X	T	R	A	O	R	D	I	N	A	R	Y	F
Y	E	N	E	R	G	E	T	I	C	A	L	L	Y	V	M	T

Misspelt words with unstressed vowels

Write the missing words in these sentences. The words you need are hidden in the grid. The letters are in the correct order but the words have been muddled together. The first one has been done as an example. Write the words on the lines like the example. The number in brackets shows how many letters are in the missing word.

H	E	L	P	C	I	P	T	L		
D	O	F	L	N	T	T	E	E	R	
C	X	M	I	I	O	T	T	I	Y	
E	E	P	I	B	I	A	I	O	O	
E	U	H	E	E	C	I	T	O	N	
D	X	P	I	D	I	T	I	E	N	N

1. I made *duplicate* copies of the important documents. (9)
2. The art ________________ was full of beautiful pictures. (10)
3. The ________________ up Mount Everest was very dangerous. (10)
4. It was a very close ________________; no one knew who would win. (11)
5. I am ________________ going swimming, even if it is cold. (10)
6. The ________________ hovered over the canyon and the view was amazing. (10)

T	E	P	V	I	C	T			
S	R	R	T	R	N	C	L		
S	E	S	O	A	A	E	E	L	
P	E	A	C	P	T	A	I	L	Y
D	R	N	G	I	L	L	N	N	Y
A	I	I	A	E	P	A	E	E	E

7. We got there just in time to see the ________________ take off. (9)
8. My dad says I should ________________ my puppy to make him behave. (10)
9. The sign on the gate said '________________ No Entry'. (7)
10. We all drove to the beach ________________, but we met up when we got there. (10)
11. My teacher gets angry when we don't begin each ________________ with a capital letter. (8)
12. The child was ________________ killed by a freak wave. (10)

Commonly misspelt words or easily confused spellings

The words in this puzzle end in **IBLE** and **ABLE**. The rule for adding these suffixes is as follows:

- **ABLE** is usually added to complete words, for example, *accept – acceptable*;
- **IBLE** is usually added to roots that are not complete words, for example, *possible*.

Here are some guidelines:

- A final **E** is usually removed before adding **ABLE** unless the word ends in a soft **G**, for example, *love – lovable* and *manage – manageable*.
- Change **Y** to **I** before adding **ABLE**, for example, *deny – deniable* and *rely – reliable*.

Some **ABLE** exceptions are: *capable, impeccable, malleable.*
Some **IBLE** exceptions are: *accessible, collapsible, flexible.*

Choose the correct suffixes for these words and write them in the right-hand column.

	Choice of spellings		Correct spelling
1	INVISABLE	INVISIBLE	
2	HORRABLE	HORRIBLE	
3	INCREDABLE	INCREDIBLE	
4	BELIEVABLE	BELIEVIBLE	
5	IMPOSSABLE	IMPOSSIBLE	
6	LAUGHABLE	LAUGHIBLE	
7	EDABLE	EDIBLE	
8	REASONABLE	REASONIBLE	

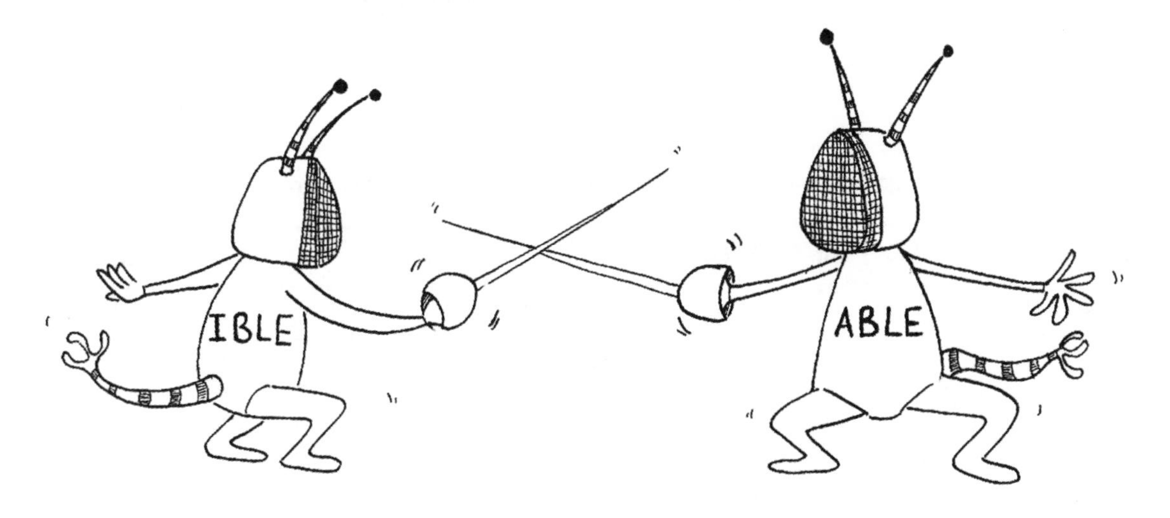

Commonly misspelt words or easily confused spellings

These verbs and nouns have related meanings and slightly different spellings because spelling changes when the word is a noun or a verb. The rules are as follows:

- **C** for a noun and **S** for a verb, for example, *advice* (noun) – *advise* (verb);
- **TH** for a noun and **THE** for a verb, for example, *breath* (noun) – *breathe* (verb);
- **F** for a noun and **VE** for a verb, for example, *belief* (noun) – *believe* (verb).

Sort these words into nouns and verbs in the table below.

advice advise bath bathe belief believe thief thieve cloth
clothe grief grieve half halve licence license relief relieve
breath breathe shelf shelve life live

Noun		Verb	

The endings **ORY**, **ERY** and **ARY** can also be confusing. Some tips to help are as follows:

- **ARY** – added to nouns or adjectives, for example, *documentary* or *legendary*;
- **ERY** – added to nouns only, for example, *bakery*, *pottery* and *machinery* (One exception is *slippery*, which is an adjective.);
- **ORY** – added to nouns or adjectives, for example, *category* or *contradictory*.

Add the correct endings to these root words and write them on the lines.

	Word	Correct spelling		Word	Correct spelling
1	annivers _ _ _		7	ordin _ _ _	
2	bound _ _ _		8	moment _ _ _	
3	deliv _ _ _		9	myst _ _ _	
4	categ _ _ _		10	nurs _ _ _	
5	comment _ _ _		11	jewell _ _ _	
6	discov _ _ _		12	necess _ _ _	

Commonly misspelt words or easily confused spellings

Words that end in **ANT** and **ENT** are usually adjectives. Some exceptions to this rule are *inhabitant, irritant, continent* and *effluent* which are all nouns. Also, when two words end in **ENT** and **ANT**, such as *dependent – dependant*, the adjective is **ENT** and the noun is **ANT**. The adjectives in the puzzle below are frequently used words that you need to learn and remember. There is also an important connection between the suffixes **ENT** and **ENCE**, and **ANT** and **ANCE**, for example, *defiant – defiance* and *different – difference*.

Add the correct root words to these ending after reading the clues.

	Clue	Ending	Complete adjective
1	Not the same.	**ENT**	
2	Nice or agreeable.	**ANT**	
3	First class, outstanding.	**ENT**	
4	Not willing to wait.	**ENT**	
5	Not guilty.	**ENT**	
6	Well behaved.	**ENT**	
7	Significant or of great consequence.	**ANT**	
8	Not keen.	**ANT**	
9	Aggressive or intense.	**ENT**	
10	Needing attention at once.	**ENT**	
11	Self-reliant.	**ENT**	
12	Far away.	**ANT**	

The endings **CEDE**, **CEED** and **SEDE** all sound the same, but **CEDE** is the most common. On a separate piece of paper, draw a table with three coloumns: '**CEDE**', '**CEED**' and '**SEDE**'. Add the words below to the table and some more of your own.

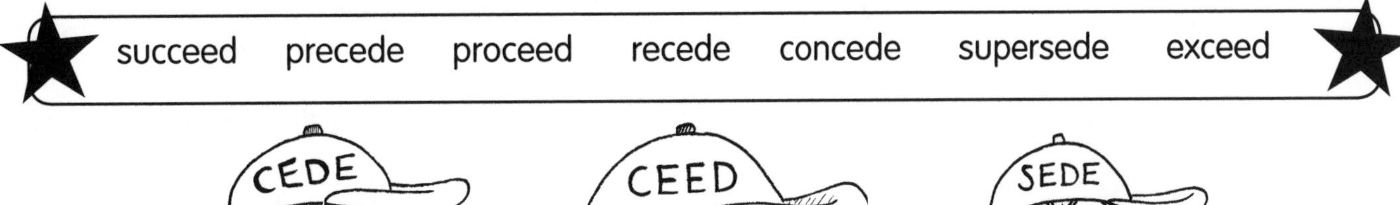

Answers

■ PAGE 6

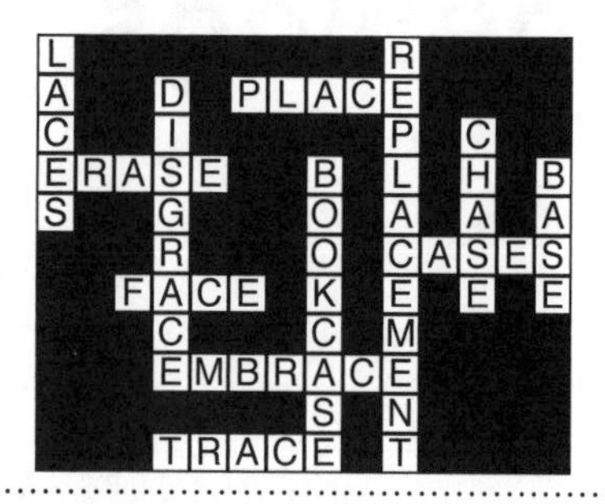

ASE and ACE – long A phoneme		ASE and ACE – short A phoneme	
deface	misplace	necklace	solace
replace	erase	furnace	palace
disgrace	staircase	terrace	purchase

■ PAGE 7

Word	Definition	Correct word
ANCHORAGE	A very wise man and an aromatic herb.	SAGE
BARRAGE	The mechanical power of a lever or influence or advantage used for a purpose.	LEVERAGE
STEERAGE	Bravery.	COURAGE
LEVERAGE	A mooring place for boats.	ANCHORAGE
HOSTAGE	To make angry.	ENRAGE
COURAGE	Part of a passenger ship with the lowest fares.	STEERAGE
WAGER	Heavy artillery fire.	BARRAGE
SAGE	A person held prisoner for ransom, or as security.	HOSTAGE
WAGES	A bet.	WAGER
ENRAGE	Payment for work done.	WAGES

1. The severe storm caused a lot of **DAMAGE** to trees and buildings.
2. The **CAGED** lions were roaring and trying to escape.
3. You usually have to pay the **POSTAGE** when you buy something by mail order.
4. The teacher said, 'This paint is very expensive so I do not want any **WASTAGE**.'
5. When you are in a foreign country you have to get used to their money which will have different notes and **COINAGE**.
6. The actors strutted around on the **STAGE**.
7. Because of gradual **LEAKAGE** the water tank ran dry.
8. If you are tall, you may have an **ADVANTAGE** when you play basketball.

■ PAGE 8

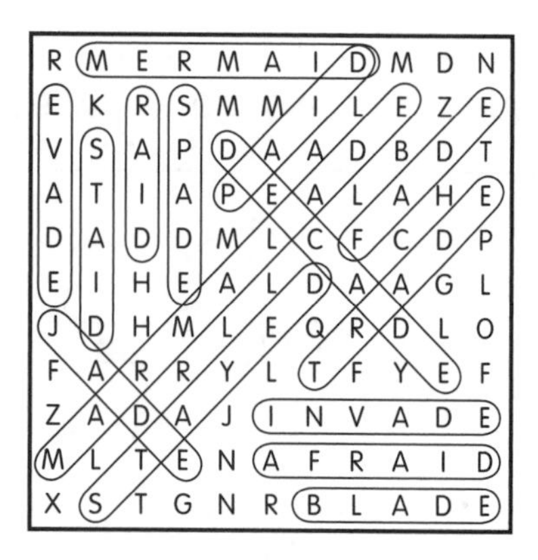

■ PAGE 9

Adjective	Verb – present tense	Verb – past tense	Noun
staid	raid	stayed	mermaid
afraid	trade	played	raid
	invade	paid	maid
	fade	inlaid	trade
	evade	prayed	spade
		decayed	jade
			decade
			blade
			marmalade

■ PAGE 10

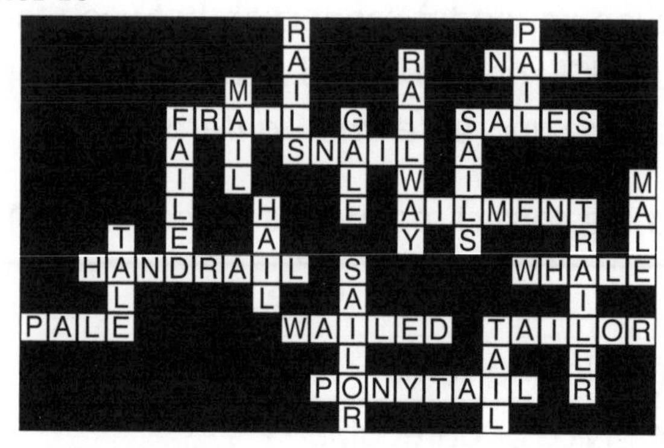

■ PAGE 11

Noun – singular		Noun – plural	Verb – present tense
ailment	male	hail	nail
tailor	nail	rails	
sailor	gale	mail	
pail	trailer		
snail	railway		
whale	ponytail	**Adjective**	**Verb – past tense**
tale	handrail	frail	failed
tail		pale	wailed
sales		male	
sails			

■ PAGE 12

1. I am going to **BAKE** a chocolate **CAKE** for tea.
2. The antique was not real; it was a **FAKE**.
3. We went rowing on the **LAKE**.
4. I made lots of **MISTAKES** in my spelling test and got lots of words wrong.
5. During the thunderstorm I couldn't get to sleep and I lay **AWAKE** in bed all night.
6. A male duck is called a **DRAKE**.
7. I heard the squeal of the car's **BRAKES** before the accident.
8. The tremors of the **EARTHQUAKE** measured 7.5 on the Richter scale.
9. A person who arranges funerals is a called an **UNDERTAKER**.
10. The pretty white **SNOWFLAKES** covered the icy ground like a lacy blanket.

Noun	Verb	Adjective	Adverb
fake	fake	shaky	shakily
Jake	bake	flaky	
cake	awake		
lake	undertake		
drake	stake		
earthquake	shake		
undertaker	quake		
snowflakes	take		
milkshake	rake		
stake	make		
flake			
baker			
rake			

■ PAGE 13

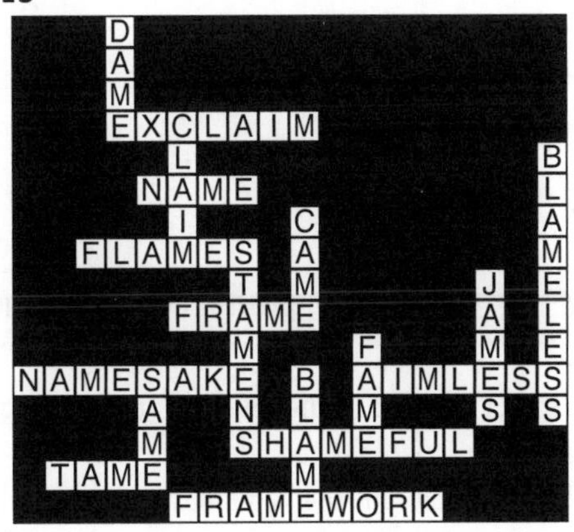

■ PAGE 14

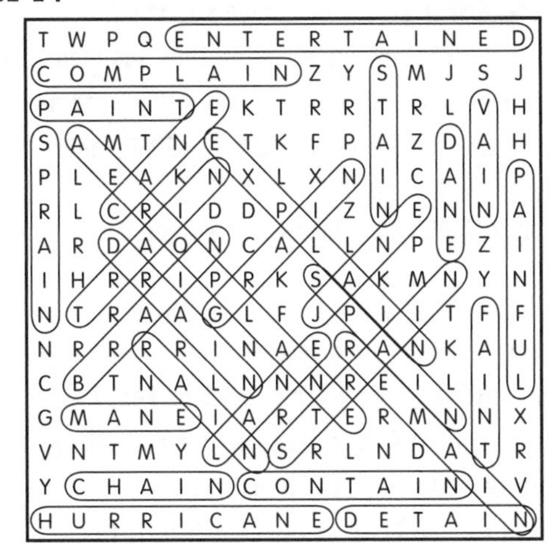

■ PAGE 15

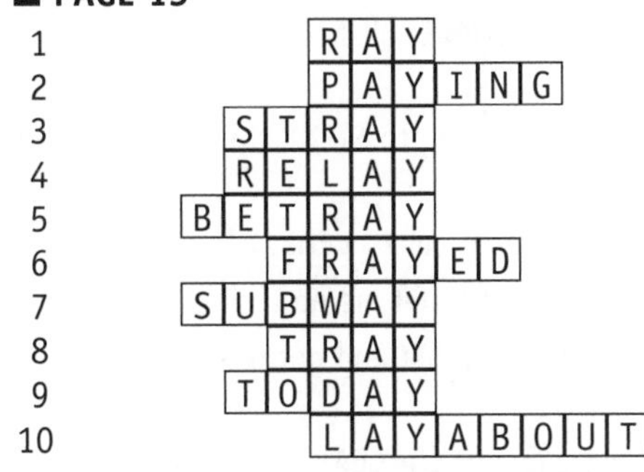

away stay yesterday castaway birthday payment holiday haywire way dismay

■ PAGE 16

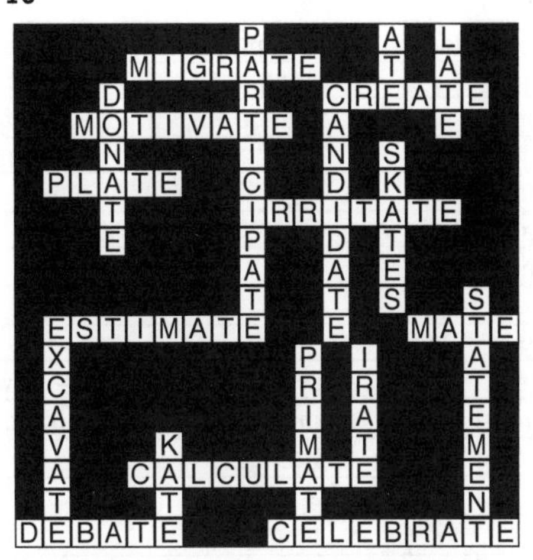

■ PAGE 17

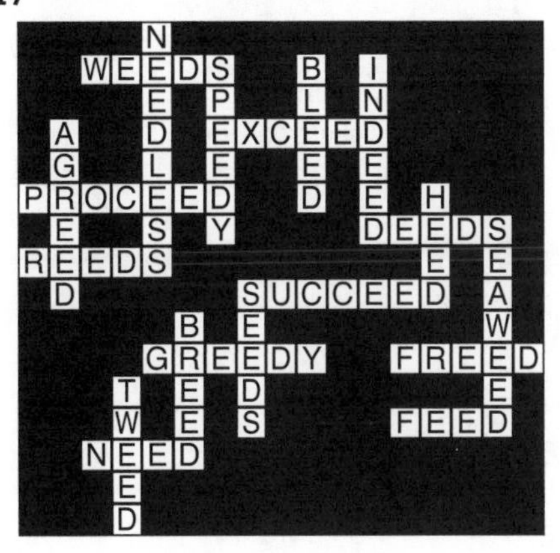

■ PAGE 18

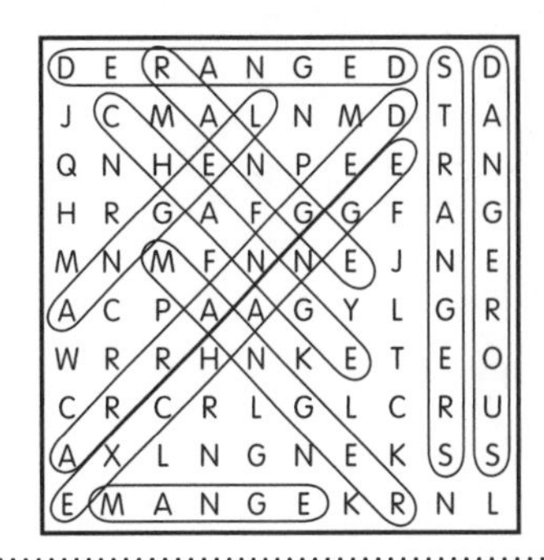

1. I **ARRANGED** the flowers in the vase and they looked beautiful.
2. The old stray dog was covered in **MANGE** and his fur was falling out.
3. In the Bible story, Jesus was placed in a **MANGER** in a stable.
4. My mum always told me not to talk to **STRANGERS.**
5. Some people love to do extreme sports which can be very **DANGEROUS.**
6. My mum said I was an **ANGEL** when I brought her breakfast in bed.
7. The mentally ill man was **DERANGED** and did not know who he was or what he was doing in hospital.
8. I like a wide **RANGE** of music, from classical to hip hop and rap.
9. At the airport I was able to **EXCHANGE** my foreign currency at a good rate.
10. The actress had to do a quick **CHANGE** behind the stage before she came on for the next act.

PAGE 19

Noun – singular	Noun – plural	Verb – present tense
mange	strangers	exchange
manger	change	change
angel		range
range		
exchange		
grange		
danger		
ranger		
Adverb	**Adjective**	**Verb – past tense**
dangerously	dangerous	arranged
strangely	strange	
	changeable	
	mangy	

PAGE 20

1. pair
2. rare
3. chair
4. hair
5. bare
6. repair
7. scare
8. nightmare
9. questionnaire
10. wear
11. there
12. mare
13. fairy
14. fare
15. lair
16. flare
17. swear

PAGE 21

1. F**AIR**Y
2. F**AIR**LY
3. SQU**ARE**
4. C**ARE**FUL
5. BEW**ARE**
6. PREP**ARE**
7. D**AIR**Y
8. WH**ERE**
9. POLAR B**EAR**
10. UNF**AIR**
11. SH**ARE**
12. FANF**ARE**
13. NOWH**ERE**
14. THOROUGHF**ARE**
15. TH**ERE**
16. SW**EAR**
17. ANYWH**ERE**
18. REP**AIR**

1. In maths, we had to **COMPARE** the fractions and work out which was the biggest.
2. The burglars were so quiet that the sleeping family were **UNAWARE** they were in the house.
3. The rabbit froze in the **GLARE** of the car's headlights.
4. I love cream cakes, especially chocolate **ECLAIRS**.
5. After winning the lottery, the man became a **MILLIONAIRE**.
6. My grandad loves to sit in his favourite **ARMCHAIR** and read the paper.
7. At the **FAIR** the children ate candyfloss and went on all the rides.
8. I didn't have enough money to pay the bus **FARE**, so I had to walk home.
9. The celebrity cut the ribbon and said 'I **DECLARE** that this sports centre is now open!'
10. The carpet was very old and **THREADBARE**, so we bought a new one.
11. At the shop I bought some **PEARS** for the fruit salad.
12. We looked **EVERYWHERE** for the missing money.

PAGE 22

1. The old hinges began to **creak** as the door moved in the wind and I could hear the water in the **creek** flowing by.
2. I climbed to the top of the mountain **peak** for a **peek** at the view.
3. I needed lots of fresh **leeks** to make my soup and I hoped the old pan didn't have any **leaks**.
4. After I was sick I felt very **weak** for over a **week**.

1. SQU**EAK**
2. FR**EAK**
3. M**EEK**
4. S**EEK**
5. SP**EAK**
6. STR**EAK**
7. WR**EAK**
8. BL**EAK**
9. GR**EEK**
10. CH**EEK**
11. SL**EEK**

1. SQUEAKY 2. SPEAKING 3. CHEEKY 4. SEEKING

PAGE 23

1. I began to **FEEL** scared as the rollercoaster went higher and higher.
2. After the marathon run, she had a nasty blister on her **HEEL**.
3. The boat's **KEEL** had a big hole in it and we began to sink.
4. We had to **KNEEL** to say our prayers in church.
5. A wanted to **PEEL** my apple, as the skin was blemished.
6. For Christmas, my brother got a new fishing rod and **REEL**.
7. The saucepans were made of stainless **STEEL** and would not tarnish.
8. The front **WHEEL** of my bike was badly dented in the accident.
9. I got a great **DEAL** when I bought my new computer and I paid much less than I expected.
10. For cuts to **HEAL** properly, they must be kept clean and dry.
11. My umbrella was **IDEAL** for keeping the sun off my face.
12. My favourite **MEAL** is pizza and chips.
13. You could hear the **PEAL** of the church bells from miles away.
14. The antique wasn't **REAL**; it was a fake.
15. To make my baby cousin **SQUEAL**, all I have to do is tickle her.
16. My mum told me it is very wrong to **STEAL** things from other people.
17. The meat from young calves is called **VEAL**.
18. I had to **APPEAL** to my mum to let me go to the party.
19. The baby **SEAL** had lost its mother and got washed up on the beach.
20. I used some touch-up paint to **CONCEAL** the scratch I had made on my dad's car.

■ PAGE 24

1. DEEPER
2. KEEP
3. SLEEP
4. STEEPLY
5. HOUSEKEEPER
6. CREEPER
7. REAP
8. SQUEAL

Noun – singular	Noun – plural	Verb – present tense	Verb – past tense	Adjective	Adverb
leap	sheep	leap	heaped	asleep	cheaply
heap	heaps	heap	peeped	deepest	deeply
creeper		weep	beeped	steep	steeply
housekeeper		sleep		deep	
sweeper		peep		cheapest	
sleep		keep		deeper	
peep		heaps			
sheep					
jeep					
beep					

■ PAGE 25

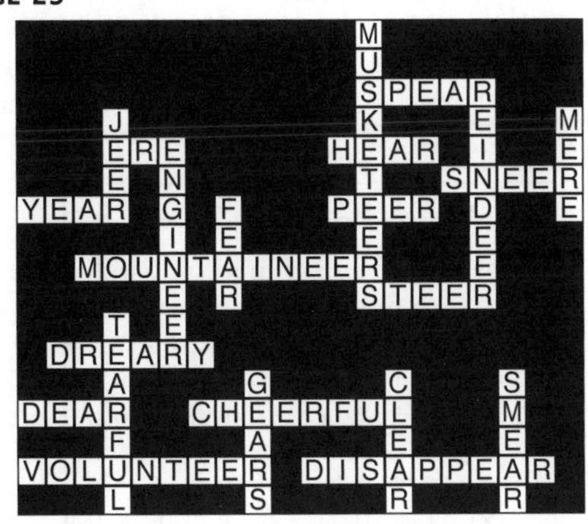

■ PAGE 26

1. CHEAT
2. CHEETAH
3. WHEAT
4. REPEAT
5. DEFEAT
6. HEAT
7. SHEET
8. SLEET
9. MEETING
10. SEAT

I am: **treatment**
I am: **fleet**

■ PAGE 27

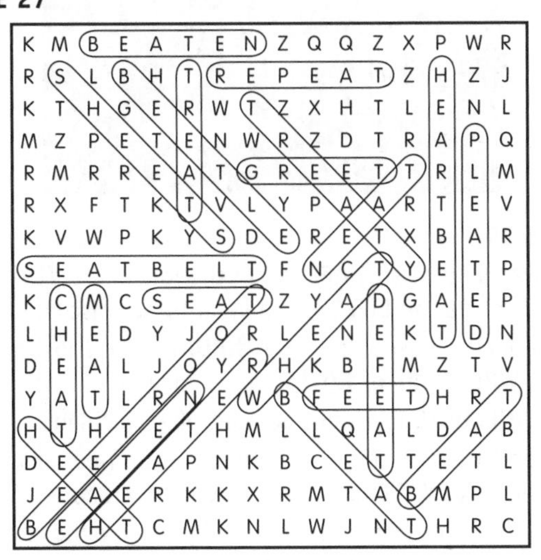

■ PAGE 28

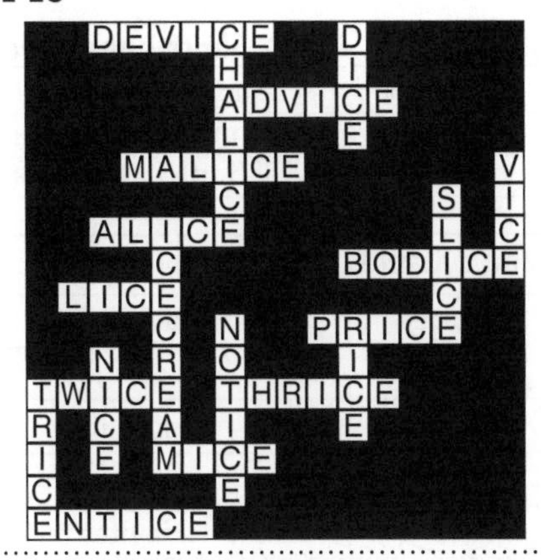

Long I phoneme		Short I phoneme
dice	trice	Alice
lice	twice	bodice
mice	advice	malice
nice	device	notice
rice	entice	chalice
vice	thrice	
price	ice cream	
slice		

■ PAGE 29

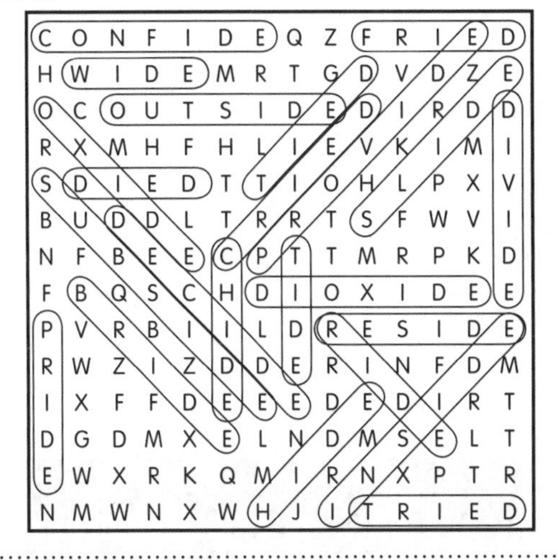

stride collide satisfied guide dried lied supplied died spied

■ PAGE 30

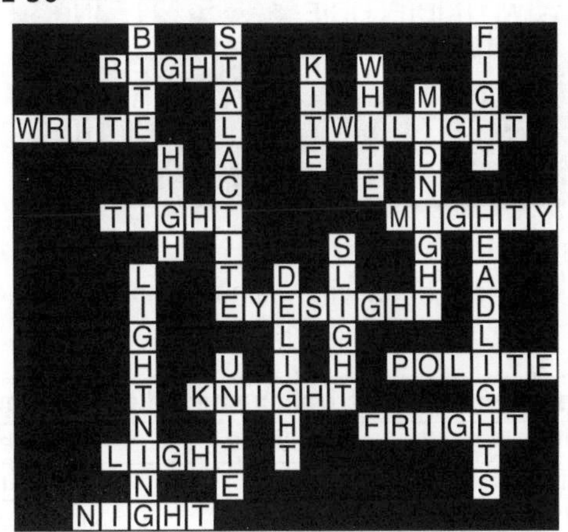

■ PAGE 31

Long I phoneme		Short I phoneme	
quite	tonight	opposite	requisite
midnight	bright	infinite	exquisite
invite	excite	definite	granite

1st syllable	2nd syllable	Complete word
spot	spite	despite
de	sight	hindsight
in	light	spotlight
air	cite	recite
hind	vite	invite
re	tight	airtight

1st syllable	2nd syllable	3rd syllable	Complete word
ap	pe	ful	delightful
co	en	ing	frightening
fright	light	right	copyright
de	py	tite	appetite

■ PAGE 32

1. TRIAL
2. FILE
3. SMILE
4. SUNDIAL
5. WHILE
6. STILE
7. EXILE
8. RECONCILE
9. CAMOMILE
10. DENIAL

1. DIALLING
2. CROCODILE
3. PROFILE
4. TRIALLED
5. DENIAL
6. SMILED

■ PAGE 33

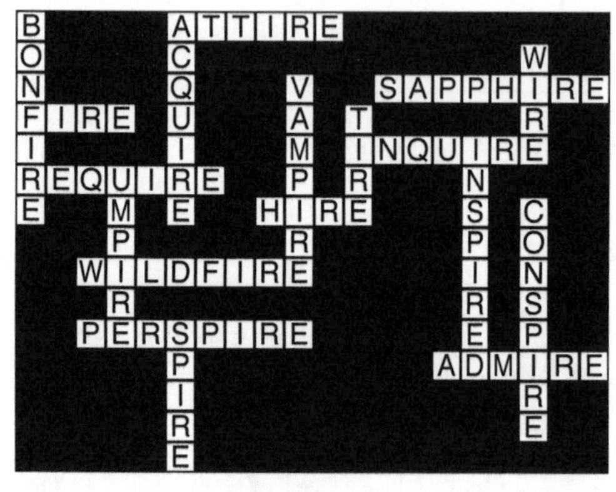

■ PAGE 34

incline outline nineteen canine spine wine pine pineapple refine line

Long I phoneme	Short I phoneme	Long E phoneme
spine	adrenaline	chlorine
combine	medicine	tambourine
incline	determine	machine
headline	feminine	gasoline
porcupine	heroine	ultramarine
airline	discipline	nectarine
recline	imagine	submarine
decline	engine	margarine

■ PAGE 35

1. UNDERLINE
2. SHRINE
3. ALKALINE
4. MINEFIELD
5. CANINE
6. SPINELESS
7. TWINE
8. NINETY
9. GOLDMINE
10. BRINE

■ PAGE 36

1st syllable OKE		2nd syllable OKE	
CHOKER	STROKE	SUNSTROKE	PROVOKE
POKER	YOKEL	AWOKEN	OUTSPOKEN
BROKEN		BACKSTROKE	STOCKBROKER
SPOKEN		EVOKE	PAWNBROKER

1. awoken 2. broken 3. spoken

1. invoke 2. provoke 3. revoke 4. evoke

1. joker 2. choker 3. stoker 4. poker 5. broker

■ PAGE 37

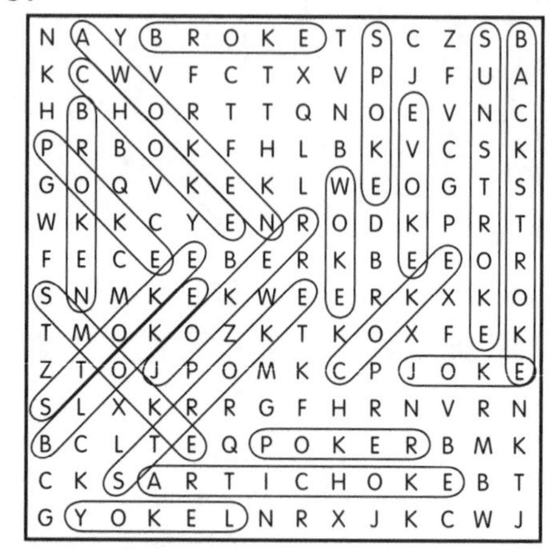

■ PAGE 38

■ PAGE 39

1. It was hard to stay **AFLOAT** in the rough sea; I was so glad to see the **LIFEBOAT** coming to our rescue.
2. The doctor looked down my **THROAT** when I lost my voice and **WROTE** lots of **NOTES** on his computer.
3. An **ANECDOTE** is a short account of an incident.
4. When I went to the boss's office I was afraid he was going to **DEMOTE** me, but luckily I got promoted instead.
5. I liked the country cottage; I didn't even mind the pet **GOAT** but it was too **REMOTE** for me as I wanted to be closer to town.
6. In the olden days, women used to wear **PETTICOATS** under their skirts.

7. After the huge five course meal I felt really **BLOATED** and I couldn't even do up my **COAT**.
8. The boy was **DEVOTED** to his mother.
9. In my English Literacy exam I tried to **QUOTE** some famous authors.
10. When I won the election with nearly all the **VOTES**, it was hard not to **GLOAT** and get big-headed.

promote vote motorboat gloat antidote
devote rote gloating noted

PAGE 40

UNDERSTOOD
WOODEN
WOOD
NEIGHBOURHOOD
MOODY
CHILDHOOD
HOODWINK
WITHSTOOD
MISUNDERSTOOD
GOOD

PAGE 41

Short OO phoneme		Long OO phoneme
withstood	wood	mood
childhood	hoodwink	broody
livelihood	stood	food
neighbourhood	misunderstood	
good		

1. CRUDE
2. LIKELIHOOD
3. MOTHERHOOD
4. INTRUDER
5. SECLUDED
6. INTERLUDE
7. KNIGHTHOOD
8. GOODNESS
9. UNDERSTOOD
10. EXCLUDED

PAGE 42

1 syllable	2 syllables	3 syllables
SOON	IMMUNE	HONEYMOON
TUNE	SCHOONER	AFTERNOON
MOON	TYPHOON	MISFORTUNE
SWOON	PLATOON	MACAROON
SPOON	MONSOON	TABLESPOON
NOON	LAGOON	DESSERTSPOON
PRUNE	HARPOONS	
DUNE	CARTOONS	
CROON	FORTUNE	
	BALLOONS	
	TEASPOON	

PAGE 43

1. At the beach I love playing in the sand **DUNES**.
3. For the party we blew up lots of **BALLOONS**.
3. A dried plum is called a **PRUNE**.
4. We lay the table and put out the knives, forks and **SPOONS**.
5. A full **MOON** appears in the sky every 28 days.
6. My mum buys lottery tickets and hopes to win a **FORTUNE**.
7. I am a terrible singer, as I can't sing in **TUNE**.
8. My little cousin loves to watch children's **CARTOONS** on the TV.
9. On holiday, we swam in the **LAGOON**.
10. It rains a lot in the **MONSOON** season.
11. A small unit of soldiers is called a **PLATOON**.
12. A **RACCOON** is a small North American animal with a bushy striped tail.
13. Another name for a very violent windy storm is a **TYPHOON**.
14. After a couple get married, they often go on a **HONEYMOON**.
15. A sailing ship with at least two masts is called a **SCHOONER**.
16. If you are **IMMUNE** to a specific illness, you will not catch it.

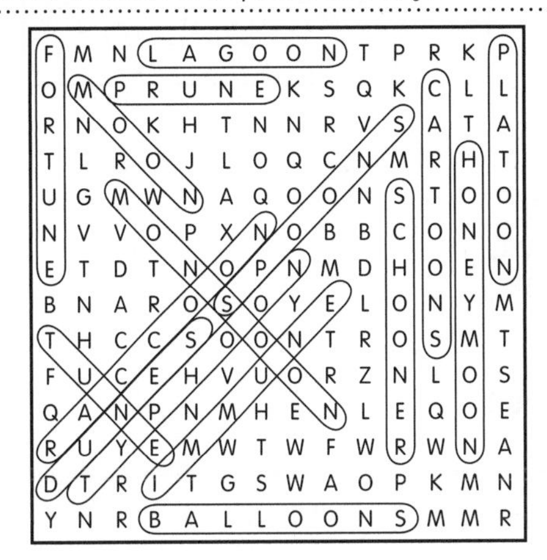

PAGE 44

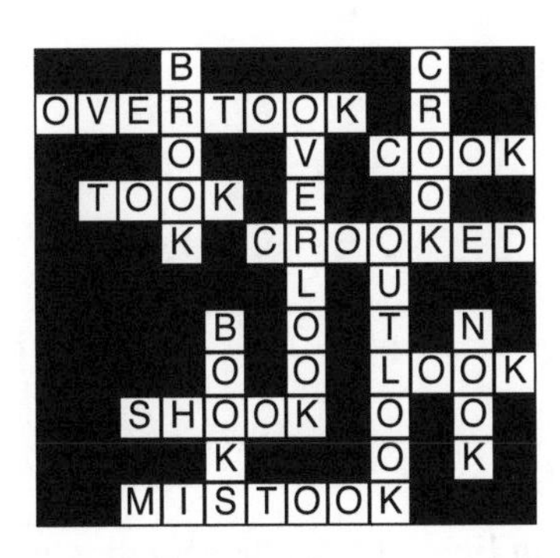

PAGE 45

1. The farmer put a **SCARECROW** in his field to frighten away the birds.
2. I weeded the garden and put all the waste in a **WHEELBARROW**.
3. I stood **BELOW** the Eiffel Tower and looked up at it in amazement.
4. **ALTHOUGH** I had won the race I wasn't pleased with my time.
5. The explorers didn't want to catch malaria so they slept inside **MOSQUITO** nets.
6. The sunflowers were a beautiful golden **YELLOW** colour.
7. My gran moved into a **BUNGALOW** so that she wouldn't have to use the stairs.
8. **TOMORROW** is the day after today.
9. At Christmas, people kiss under the **MISTLETOE**.
10. Country lanes can be really **NARROW** and not wide enough for two cars.
11. When my brother ran out of money, he tried to **BORROW** some from my mum.
12. I heard an **ECHO** as I called my name out in the cave.
13. When the avalanche began, the mountaineers tried to **RADIO** for help.
14. It was still raining but the sun came out and we saw a **RAINBOW**.
15. When I kicked the ball, it went straight through the open **WINDOW**!

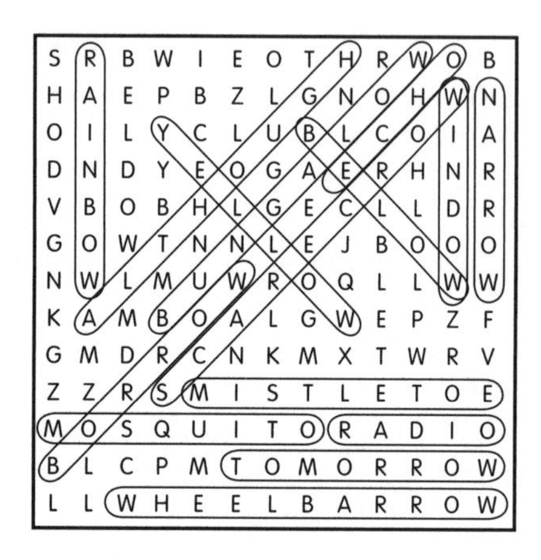

■ PAGE 46

1. DOCT**OR**
2. BEF**ORE**
3. PINAF**ORE**
4. HORR**OR**
5. W**ORE**
6. MAJ**OR**
7. GLAM**OUR**
8. F**OUR**TEEN
9. IGN**ORE**
10. NEIGHB**OUR**
11. SPONS**ORE**D
12. TERR**OR**
13. C**OUR**T
14. IMPL**ORE**
15. MAN**OR**
16. M**OUR**NFUL
17. MAY**OR**
18. F**OR**TY
19. ENDEAV**OUR**
20. SAIL**OR**

■ PAGE 47

ORE sounding like OR		ORE sounding more like ER	
for	afford	minor	anchor
forty	wore	major	endeavor
fourth	yore	bachelor	horror
course	pork	conductor	manor
court	cork	doctor	neighbour
mournful	fork	honour	glamour
your	sore	armour	work
pour	implore	world	worm
bore	ignore	terror	sponsored
storm	normal	sailor	comfort
torn	before	scissors	worth
adore	horn	effort	sculptor
port	sport	protestor	tremor

1 syllable	2 syllables	3 syllables	4 syllables	5 syllables
wore	afford	important	ordinary	evaporation
course	honour	organise	predecessor	commemoration
sore	adore	endeavour	disorganised	organisation
pour	armour	conductor	carnivorous	territorial
sport	sailor	professor	exploration	dishonourable

■ PAGE 48

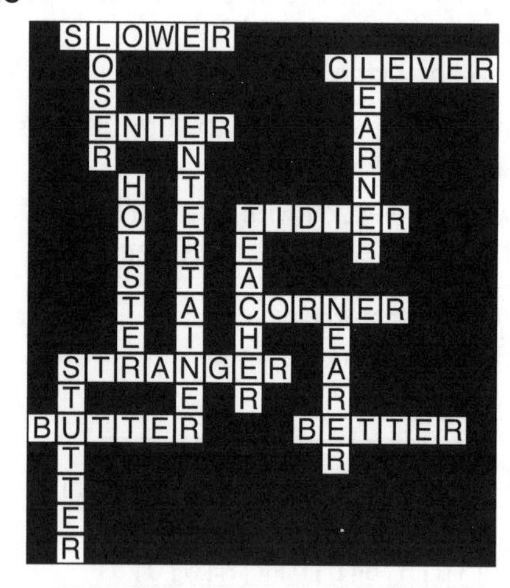

■ PAGE 49

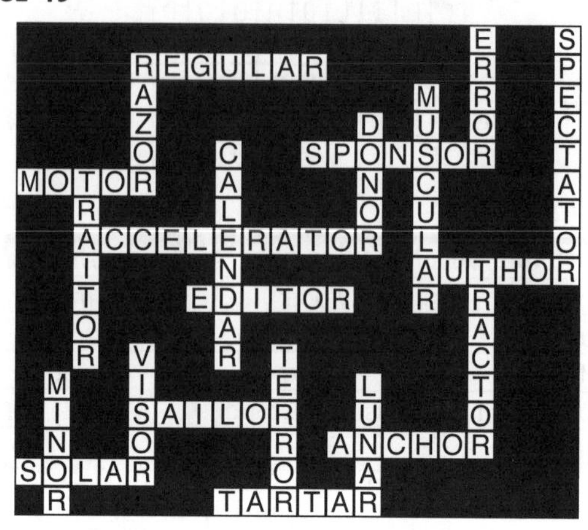

■ PAGE 50

centigrade awake mundane illustrate behave amazement exhale lace

centipede theme scene these complete trapeze evening precede

advice divide landslide crocodile grime combine exercise capsize

explode gnome tadpole postpone telescope closed cove frozen

introduce huge duke molecule perfume tuneful picture injure

■ PAGE 51

1. CONCENTRATE
2. DECADE
3. PROMOTE
4. CONSUME
5. ESCAPE
6. PRECEDE
7. ANTIDOTE
8. HYGIENE
9. COMPETE
10. TWICE
11. JUVENILE
12. OPPOSE
13. PERSPIRE
14. DELUGE
15. CONCAVE

1. marmalade
2. earthquake
3. volume
4. stethoscope
5. sapphire
6. civilise
7. female
8. ridicule
9. cyclone
10. surname
11. hurricane
12. adventure
13. Japanese
14. paradise
15. tessellate
16. grindstone
17. overtime
18. stalagmite
19. dispose
20. prosecute

■ PAGE 52

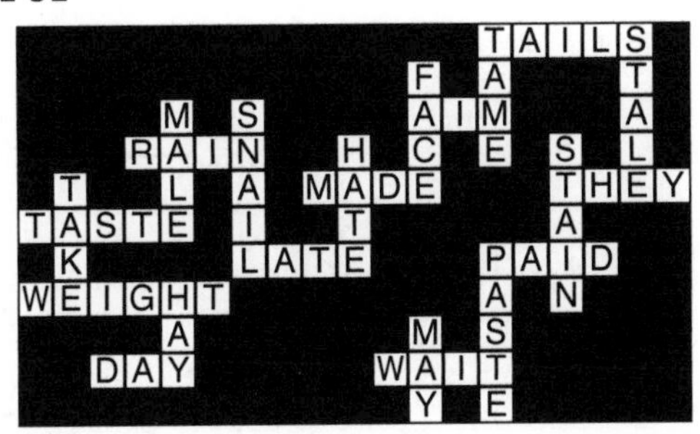

■ PAGE 53

A_E(or A_E root word)	AI	A + single consonant	AY
vertebrate	waist	apron	display
communication	complaint	April	mislay
alternating	painting	famous	displayed
location	again	alien	Wednesday
vacation	explain	agent	playing
insulate	mermaid	raven	payment
pollinate	faint	Adrian	away
illustrate	**EY**	Asia	delay
danger	obeyed	nature	Norway
flame	they	Australia	everyday
waste	survey		straying
label	disobeying		subway
misplace	surveyor		portray
bookcase			
phrase			
opaque	**LE ending**	**EI**	
landscape	gable	reins	
nightingale	stable	reign	
masquerade	cradle	eight	
decade	ladle	sleigh	
spacious	table	veil	
lazy			
fatal			

■ PAGE 54

1. DISAGR**EE**
2. P**EA**CE
3. P**IE**CE
4. CENTIP**E**D**E**
5. Z**E**RO
6. S**EA**W**EE**D
7. **EA**GLE
8. SUPR**E**M**E**
9. INCR**EA**SE
10. BR**IE**F
11. PION**EE**R
12. **E**QUATOR
13. CONC**EA**L
14. CONC**EI**TED
15. D**EE**PER

■ PAGE 55

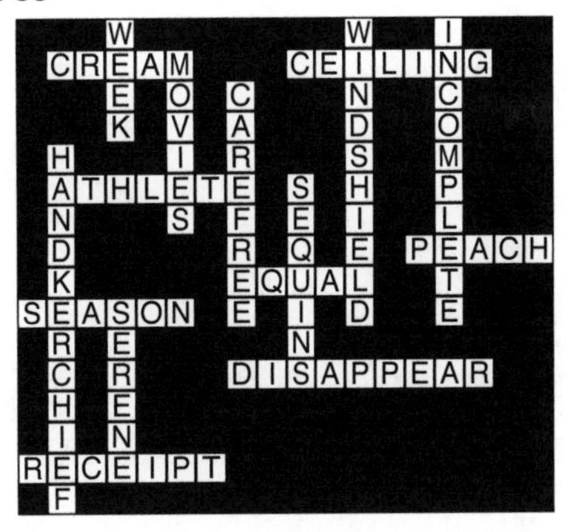

■ PAGE 56

1. The bride walked down the aisle of the church, wearing a light blue bridal gown.
2. The wild animals were sleeping in the shade, as we tried to spy on them for a while.
3. At night they went hunting and travelled for miles across different tribal lands.
4. My little sister has a tricycle with wide tyres and I have a bicycle with nice shiny paintwork.
5. The man replied to the policeman's questions and denied all knowledge of the crimes.
6. We ignited a big bonfire and the flames shot high into the night sky, while the sparks began to fly like shooting stars.
7. I didn't have time to hide the prizes right away and the children started to fight over them.
8. He would have been released, but he was unwise and tried to bribe the police. They soon found out that he had lied.
9. The cowboy put the bridle on his ride, loaded his rifle, shined his grimy boots and galloped away, just in time.
10. The mountain was so high, we climbed for nine hours. My throat was dry and my thighs were tired.

■ PAGE 57

Noun	Noun – plural	Verb – present tense	Verb – past tense	Adverb	Adjective
fire	tiles	bribe	satisfied	kindly	grimy
isle	miles	fire	lied	nicely	slimy
spine	thighs	shine	surprised	blindly	mild
wife	sighs	unwind	dried	tightly	childish
pie	tyres	sighs	denied	shyly	tidal
supplies	rifles	fight	ignited	childishly	higher
kite	prizes	invite	tired	mildly	spicy
fight	wives	apply	applied	excitedly	blind
idol	pies	deny	reminded	lightly	white
reply	spices	satisfy		wisely	shy
sighs		spy			ladylike
trifle		reply			idle
sky		spices			wise
spy					

PAGE 58

1. home
2. over
3. own
4. road
5. toe
6. Polish
7. comb
8. go
9. shoulder
10. bowl
11. woven
12. cone
13. oval
14. drove

PAGE 59

1st syllable	2nd syllable	3rd syllable
broken	wardrobe	tobacco
hopeless	postpone	episode
postpone	soapstone	inferno
soapstone	oboe	Apollo
overdose	windows	overdose
oboe	solo	interrogate
solo	photograph	indigo
photograph	October	mistletoe
opponent	chocolate	dynamo

1. COAT
2. ROSES
3. FOLD
4. LINO
5. ELBOWS
6. SLOWCOACH
7. HOLES
8. CROAK
9. GOAL
10. ALONE
11. ROWING
12. FLOAT
13. BACKBONE
14. BOAST
15. INDIGO

PAGE 60

~~R~~	[4]O	~~F~~	~~U~~	[2]G	~~E~~	~~D~~		
[3]S	[3]O	[4]R	[2]U	[1]U	[4]N	[4]A		
[2]D	~~E~~	[3]L	[3]U	~~S~~	[2]E	[3]E		
[1]P	[1]E	[1]R	[4]T	[3]B	[1]M	[1]E		
[4]F	[2]E	[2]L	[1]F	[4]U	[3]L	E	[4]T	[4]E

1. PERFUME
2. DELUGE
3. SOLUBLE
4. FORTUNATE

1. cue → A long stick for playing snooker.
   queue → A line of people.

2. due → Expected
   dew → Tiny drops of water that form outside overnight.

3. new → Not old.
   knew → Was aware of or had information about.

PAGE 61

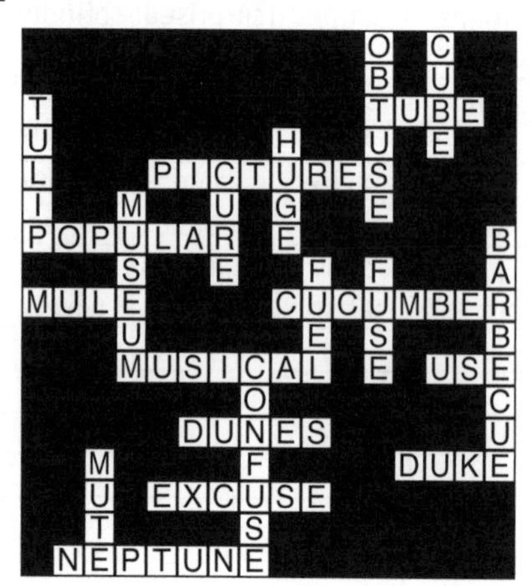

PAGE 62

	Singular form	Plural form		Singular form	Plural form
★	fox	foxes	★	scratch	scratches
★	toy	toys	★	baby	babies
★	lunch	lunches	★	brush	brushes
★	month	months	★	friend	friends
★	spy	spies	★	display	displays
★	half	halves	★	scarf	scarves
★	ash	ashes	★	pie	pies
★	cough	coughs	★	leaf	leaves
★	fly	flies	★	toffee	toffees
★	wave	waves	★	cave	caves

PAGE 63

	Singular form	Plural form		Singular form	Plural form
★	man	men	★	gas	gases
★	apple	apples	★	shelf	shelves
★	chief	chiefs	★	woman	women
★	person	people	★	child	children
★	ear	ears	★	potato	potatoes
★	loaf	loaves	★	church	churches
★	foot	feet	★	tooth	teeth
★	piano	pianos	★	table	tables
★	dish	dishes	★	hoax	hoaxes
★	fish	fish	★	cactus	cacti

~~Y~~	[4]H	~~L~~	~~L~~					
[1]A	[2]E	[1]O	[2]O	[1]O	~~Y~~			
[2]H	[3]I	[4]I	[3]E	[4]N	[2]S	~~S~~	[4]S	
[3]L	~~A~~	[2]R	[1]L	~~E~~	[1]G	[4]Y	[1]E	
[4]C	[1]P	[3]C	[4]M	[2]E	[4]E	[1]I	S	[1]S

1. APOLOGIES
2. HEROES
3. LICE
4. CHIMNEYS

[5]T	[6]E	[5]M	[9]E	[6]E					
[6]G	[5]O	[8]U	[7]E	[9]Y	[6]E				
[7]B	[9]L	[9]L	[8]R	[7]S	[5]O				
[8]Y	[7]U	[7]S	[6]S	[8]S	[9]S	[8]L	[5]S		
[9]A	[8]O	[6]E	[5]A	[5]T	[8]E	[5]E	[8]V	[8]E	[8]S

5. TOMATOES
6. GEESE
7. BUSES
8. YOURSELVES
9. ALLEYS

PAGE 64

Verb	Verb with ED	Verb with ING
wait	waited	waiting
grumble	grumbled	grumbling
happen	happened	happening
borrow	borrowed	borrowing
dry	dried	drying
explode	exploded	exploding
stray	strayed	straying
scream	screamed	screaming
scratch	scratched	scratching
enjoy	enjoyed	enjoying
illustrate	illustrated	illustrating
stir	stirred	stirring
tour	toured	touring

PAGE 65

Hard C	Hard G	Soft C	Soft G
catapult	goalposts	certainly	generally
collar	gasping	Cyprus	giraffe
curtains	goose	Cinderella	fragile
magical	galleon	special	large
ridiculous	goggles	spaceman	dangerous
local	bigger	magician	gymnastics
pelican	regularly	politician	original
occurring	bingo	recieve	imagination

PAGE 66

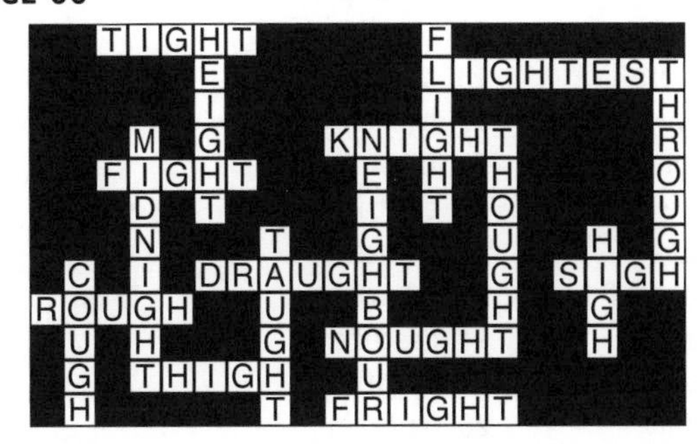

PAGE 67

Long E		Long A	Long I
conceited	perceive	veil	eiderdown
deceit	receive	reindeer	Fahrenheit
achieve	seize	reign	height
belief	weir	heir	poltergiest
masterpiece	weird	freight	seismograph
retriever	yield	vein	kaleidoscope
series	protein	sleigh	
siege	shriek	weigh	
grieve	brief	neighbour	
hygienic	niece	weight	
pierce	fiend	reins	

1. HANDKERCHIEF
2. RECEIPT
3. SLEIGH
4. PIERCE
5. ACHIEVEMENT
6. FREIGHTER
7. FRIEND
8. REPRIEVE

PAGE 68

Suffix	Noun	Verb
ISE	idol	idolise
	victim	victimise
ATE	liquid	liquidate
	migrant	migrate
IFY	class	classify
	clear	clarify
EN	fright	frighten
	wide	widen

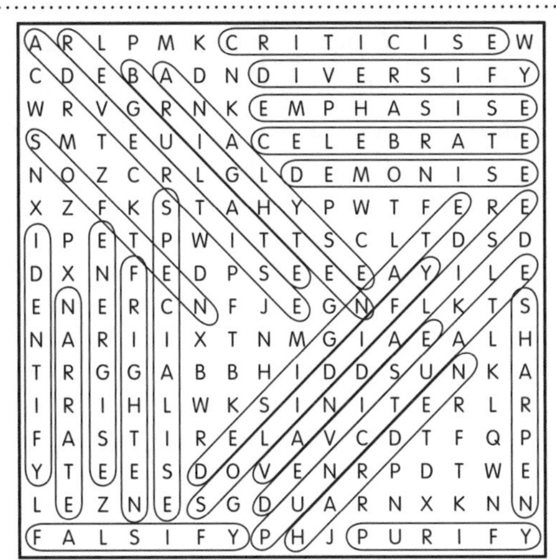

PAGE 69

Suffix	Verb	Noun
ADE	block	blockade
AGE	block	blockage
AL	refuse	refusal
ANCE	assist	assistance
ANCY	occupy	occupancy
ANT	occupy	occupant
AR	beg	beggar
DOM	free	freedom
EE	employ	employee
ENT	correspond	correspondent
ER	teach	teacher
ION	infect	infection
MENT	astonish	astonishment
OR	direct	director

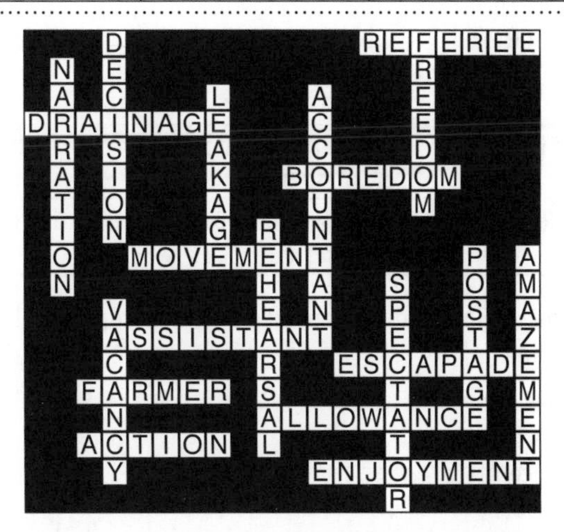

PAGE 70

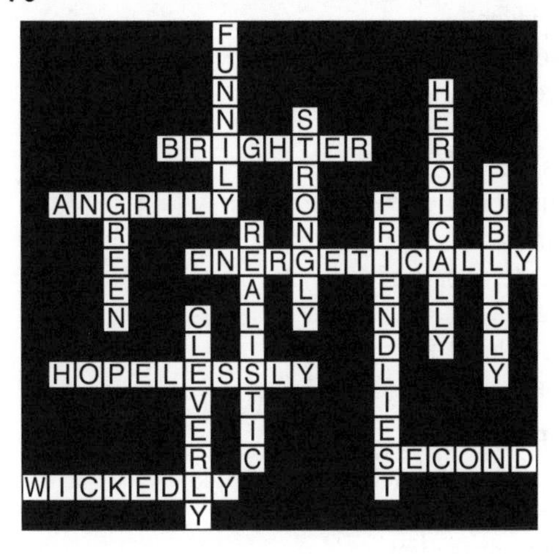

■ PAGE 71

Word	Definition	Correct word
MONOLOGUE	To set out rules, or order the use of certain medicines.	PRESCRIBE
PRESCRIBE	A study or an account of a family tree.	GENEALOGY
EPILOGUE	An instrument used in a submarine or to give views of things on different levels.	PERISCOPE
INCLUDE	A long speech by one person.	MONOLOGUE
GENEALOGY	To say what someone or something is like.	DESCRIBE
PERISCOPE	A closing speech.	EPILOGUE
TRILOGY	A series of three related works or stories.	TRILOGY
EXCLUDE	To involve, add in or make part of.	INCLUDE
DESCRIBE	A tube that you look through to see brightly coloured changing patterns.	KALEIDOSCOPE
KALEIDOSCOPE	To come or go before or in front of something.	PRECEDE
PRECEDE	To shut or keep someone or something out.	EXCLUDE

dialogue intercede conclude stethoscope eulogy transcribe

■ PAGE 72

1. comfortable
2. concentrate
3. terrible
4. conclude or concede
5. forbidden
6. beautify
7. historian
8. reversal
9. cassette
10. voluntary
11. politician
12. precede or preclude
13. cheerful
14. brotherhood
15. lioness

■ PAGE 73

1. metabolism or metabolise
2. trilogy
3. realise or realism
4. catalogue
5. decision
6. faultless
7. ladylike
8. amusement
9. faithfully
10. chronology
11. aggressiveness
12. inscribe
13. kinship
14. praiseworthy
15. telescope

1. renewable
2. supervise
3. generate
4. eulogy
5. flexible
6. exercise
7. exclude
8. epilogue
9. frozen
10. explanation
11. magnify
12. bottomless
13. equestrian
14. lifelike
15. universal
16. astonishment
17. gazette
18. thoughtfully
19. boundary
20. theology
21. electrician
22. shyness
23. intercede
24. describe
25. doubtful
26. hardship
27. childhood
28. blameworthy
29. waitress
30. microscope

■ PAGE 74

Word	Definition
AEROBATICS	Stunts carried out in the air.
CONCEAL	To hide.
CREDENTIALS	Authorisation or someone's personal details or qualifications.
AEROBICS	Exercises that strengthen your heart and lungs.
HYDROELECTRICITY	Power created by water.
AUDIOVISUAL	Information delivered with sound and pictures.
CONTRADICT	To say the opposite or that someone or some information is wrong.
CONCLUDE	To end.
AUDITION	A test to see if a performer is suitable.
CREDITABLE	Trustworthy and honest.
AQUAMARINE	A bluish-green colour.
CONTRARY	Obstinate or awkward, deliberately doing the opposite.
DUEL	A fight between two people.
AQUARIUS	A sign of the zodiac – the water bearer.
DUPLICITY	Double dealing and dishonesty.
HYDRANT	A connection for attaching a hose to the water mains.

PAGE 75

Word	Definition
OCTAVE	The range of eight notes.
PHOTOFINISH	A very close finish that needs a camera to help the decision.
SUBTERRANEAN	Underground.
OCTET	A group of eight musicians.
MICROBE	An extremely small organism too small for the eye to see.
PORTER	Someone who looks after a hotel lobby or carries guests' bags.
SUBMERGE	To put under the surface of liquid.
EXPEDITE	To hurry or speed up a process.
MICROWAVE	A cooking device that uses energy in very short waves.
PORTHOLE	A small window in the side of a ship.
TRILINGUAL	Able to speak three languages.
HYDROFOIL	A light vessel with a hull that raises out of the water at speed.
PHOTOSYNTHESIS	The process by which plants turn CO2 and water into energy.
MICROSCOPE	An instrument for magnifying minute objects.
INTERIOR	The outside.
TRIANGLE	A 2-D shape with three sides.

PAGE 76

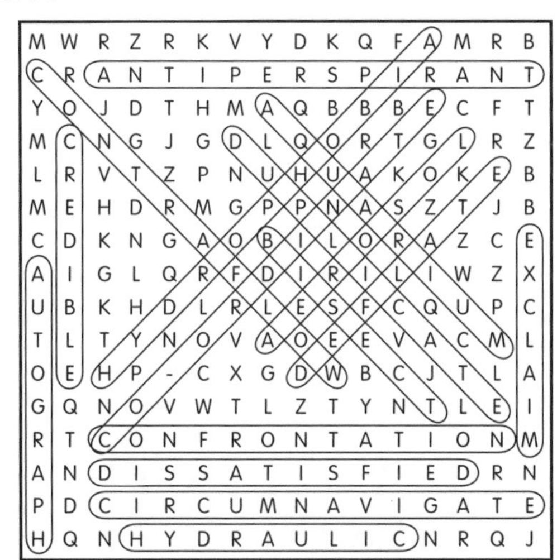

PAGE 77

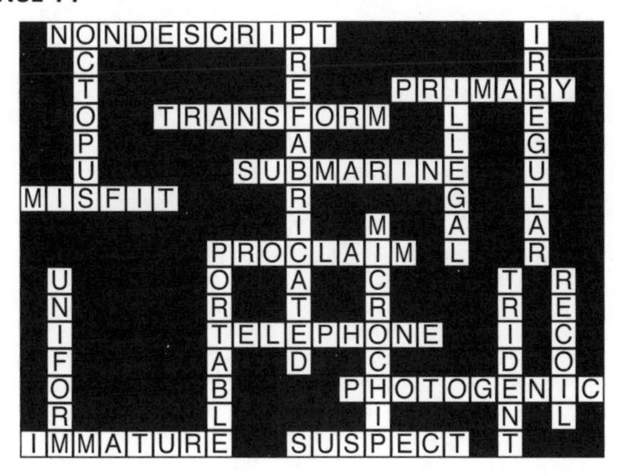

PAGE 78

1. preferable
2. mathematics
3. reference
4. desperate
5. valuable
6. fashionable
7. restaurant
8. temperament
9. cemetery
10. temporary
11. laboratory
12. veterinary
13. library
14. magically

PAGE 79

N	G	T	F	P	R	O	B	A	B	L	Y	P	C	G	X	P
E	P	E	T	W	R	H	T	B	P	L	B	M	G	J	W	K
N	Z	M	M	N	E	F	A	M	U	G	M	E	C	T	J	V
T	M	P	M	Z	V	L	V	P	R	S	L	R	E	L	E	V
H	I	O	I	R	E	L	T	C	P	B	I	N	Z	R	R	E
U	N	R	S	F	R	V	L	O	A	E	O	N	U	V	L	T
S	I	A	E	F	Y	M	O	T	W	E	N	T	E	B	J	W
I	A	R	R	R	W	N	E	L	M	E	A	I	A	S	V	Z
A	T	Y	A	B	H	G	Y	O	U	R	L	N	N	L	S	N
S	U	G	B	K	E	X	S	R	E	N	O	K	J	G	J	K
T	R	D	L	V	R	K	T	P	R	I	T	Z	Z	W	P	M
I	E	X	E	R	E	T	M	Z	H	K	R	A	K	G	N	K
C	D	C	G	W	T	E	W	S	Z	Y	P	N	R	H	T	T
A	M	F	F	P	T	K	A	K	N	K	R	N	T	Y	Q	P
L	D	D	J	F	Y	F	J	E	W	E	L	L	E	R	Y	Z
L	L	K	E	X	T	R	A	O	R	D	I	N	A	R	Y	F
Y	E	N	E	R	G	E	T	I	C	A	L	L	Y	V	M	T

PAGE 80

^{6}H	^{5}E	^{6}L	^{4}P	^{6}C	^{5}I	^{6}P	^{6}T	^{5}L		
^{5}D	^{4}O	^{5}F	^{1}L	^{5}N	^{4}T	^{5}T	^{5}E	^{6}E	^{6}R	
^{4}C	^{3}X	^{4}M	^{6}I	^{1}I	^{6}O	^{3}T	^{4}T	^{4}I	^{5}Y	
^{3}E	^{6}E	^{1}P	^{5}I	^{2}B	^{2}I	^{1}A	^{2}I	^{3}O	^{4}O	
^{2}E	^{1}U	^{2}H	^{3}E	^{4}E	^{1}C	^{4}I	^{1}T	^{2}O	^{3}N	
^{1}D	^{2}X	^{3}P	^{2}I	^{3}D	^{3}I	^{2}T	^{3}I	^{1}E	^{2}N	^{4}N

1. I made **duplicate** copies of the important documents.
2. The art **exhibition** was full of beautiful pictures.
3. The **expedition** up Mount Everest was very dangerous.
4. It was a very close **competition**; no one knew who would win.
5. I am **definitely** going swimming, even if it is cold.
6. The **helicopter** hovered over the canyon and the view was amazing.

^{12}T	^{7}E	^{10}P	^{9}V	^{12}I	^{12}C	^{10}T			
^{11}S	^{12}R	^{7}R	^{11}T	^{10}R	^{11}N	^{11}C	^{12}L		
^{10}S	^{11}E	^{8}S	^{7}O	^{9}A	^{10}A	^{9}E	^{10}E	^{12}L	
^{9}P	^{10}E	^{12}A	^{8}C	^{7}P	^{9}T	^{7}A	^{8}I	^{10}L	^{12}Y
^{8}D	^{9}R	^{11}N	^{12}G	^{8}I	^{7}L	^{8}L	^{7}N	^{8}N	^{10}Y
^{7}A	^{8}I	^{9}I	^{10}A	^{11}E	^{8}P	^{12}A	^{11}E	^{7}E	^{8}E

7. We got there just in time to see the **aeroplane** take off.
8. My dad says I should **discipline** my puppy to make him behave.
9. The sign on the gate said '**Private** No Entry'.
10. We all drove to the beach **separately**, but we met up when we got there.
11. My teacher gets angry when we don't begin each **sentence** with a capital letter.
12. The child was **tragically** killed by a freak wave.

PAGE 81

1. INVISIBLE
2. HORRIBLE
3. INCREDIBLE
4. BELIEVEABLE
5. IMPOSSIBLE
6. LAUGHABLE
7. EDIBLE
8. REASONABLE

PAGE 82

Noun		Verb	
life	licence	live	license
shelf	half	shelve	halve
breath	grief	breathe	grieve
relief	cloth	relieve	clothe
thief	belief	thieve	believe
bath	advice	bathe	advise

1. anniversary
2. boundary
3. delivery
4. category
5. commentary
6. discovery
7. ordinary
8. momentary
9. mystery
10. nursery
11. jewellery
12. necessary

PAGE 83

1. different
2. pleasant
3. excellent
4. impatient
5. innocent
6. obedient
7. important
8. reluctant
9. violent
10. urgent
11. independent
12. distant

CEDE	CEED	SEDE
precede	succeed	supersede
recede	proceed	
concede	exceed	